THE SEAS

An introduction to the study of life in the sea

SIR FREDERICK S. RUSSELL

C.B.E., LL.D., F.R.S.

*Formerly Director of the Plymouth Laboratory of
the Marine Biological Association*

AND

SIR MAURICE YONGE

C.B.E., D.Sc., F.R.S.

Honorary Fellow, University of Edinburgh

*Illustrated with 48 plates, including 24
in colour and many text drawings by
Peter Stebbing and others*

FREDERICK WARNE

PUBLISHED BY
FREDERICK WARNE & CO LTD: LONDON
FREDERICK WARNE & CO INC: NEW YORK
1975

FIRST PUBLISHED BY
FREDERICK WARNE & CO LTD
1928
SECOND EDITION 1936
THIRD EDITION 1963

© FOURTH EDITION, COMPLETELY REVISED, EXTENDED AND RESET,
FREDERICK WARNE & CO LTD, 1975

To E. J. Allen, 1866–1942

LIBRARY OF CONGRESS CATALOG CARD NO. 74-80621

ISBN 0 7232 1763 7

PRINTED IN GREAT BRITAIN BY
BUTLER & TANNER LTD
FROME AND LONDON

CONTENTS

LIST OF PLATES

*(colour plate)

PREFACE TO FOURTH EDITION

This book originated in 1927 and was published the following year. We as authors were then almost fifty years younger than we are now and marine science was notably less sophisticated. But, despite youth and limited experience, our interests were surprisingly complementary, the one concerned with plankton, fish and fisheries, the other with shore and bottom living invertebrates, with feeding mechanisms and marine cultivation. Our interests and experience covered most of the subject-matter of the book. But although we had worked around the coasts of Great Britain and Norway, and in the Mediterranean, we had no knowledge of the tropics so that the chapter on coral reefs owed nothing to first-hand experience. However, no sooner was the book published than we were both of us away to Australia, leader and deputy leader of the Great Barrier Reef Expedition which was to spend over a year on a small coral formation studying every aspect of life on coral reefs. Horizons dramatically widened and that chapter in the book was revised at the first opportunity. In other respects *The Seas* was kept up to date in successive editions and reprintings.

Few departments of knowledge have changed more over past years than those concerned with the sea. New and sometimes exciting species such as the coelocanth and the primitive molluscan *Neopilina* have been discovered. Much has been learnt about how marine animals function, how they react to their environment and to each other, how and where they reproduce themselves. The seas have become an increasingly important source of food. Some animals have been too intensely pursued, such as the Blue Whale now so reduced in numbers that the survival of the species is endangered. New fisheries have appeared, notably that for anchovies in the Humbolt Current off Peru. New methods of cultivation both of invertebrate molluscs and crustaceans and of fishes have been developed.

The invention of the apparatus for scuba diving with the development of underwater colour photography have together revealed a new world of beauty in the underwater life particularly of coral reefs. Dolphins and killer whales have been tamed to display a fascinating array of intelligence and agility in oceanaria.

The world fleet of research vessels now exceeds one thousand and marine laboratories, first established in the later decades of the last century, have increased in size and proliferated in numbers. Their work has become increasingly significant, not merely for the advance in knowledge about marine life generally, but because some marine animals have provided the means of increasing basic information. Work on squids has advanced knowledge about the nature of nerve transmission, and it is very doubtful whether this could have been obtained from other sources.

For this new edition, *The Seas* has been completely reset in a larger format. This has enabled us to alter freely and approach our subject-matter without undue reference to what we had formerly written. The original plan has been retained, although emphasis has altered and new information has been introduced. Two chapters have been added, one on the sensory perception in some marine animals—the other on the influence of man on marine life. The former chapters on the Shellfish Industry and on Products from the Sea have been combined into one dealing with Marine Exploitation and Cultivation.

Illustrations have been greatly improved. Figures in the text have been redrawn with some additions; while the plates, although inevitably reduced in number, gain from the larger page size and better methods of reproduction. For many figures and plates we are deeply indebted to Peter Stebbing who has gone to the greatest pains to check the accuracy of every illustration and has produced some very beautiful colour plates. This new enlarged edition owes much to him. Other illustrations are by Annabel Milne and Peter Berry.

The Preface of the first edition consisted largely of acknowledgements to individuals and scientific bodies who gave permission to reproduce illustrations. Their repetition is now unnecessary, but all are acknowledged in the captions and we are grateful to them. We are glad to retain some of the plates reproduced from classic reports on marine animals such as the *Fauna and Flora of the Gulf of Naples* and the *Valdivia* Reports. We are particularly pleased to reproduce again many photographs by our mutual friend Douglas Wilson to whose skill the original edition was so indebted.

Finally we thank our publishers who accepted the risk so many years ago of producing a book by two young marine biologists, then unknown to them. We hope, at least for their sakes, that this largely new edition will be as successful as were its predecessors. Over the years it has been so pleasing to be told by so many people in so many countries how much their interest in marine science was aroused by reading *The Seas*.

F. S. RUSSELL
C. M. YONGE

Chapter 1

GENERAL INTRODUCTION

The Oceans

'Oceanus', son of Heaven and Earth, was the name given by the Greeks in days gone by to an ever-flowing river that they supposed to flow around the earth they thought was flat. Later the name became applied to those waters that were far outside the range of land, being first used to signify the Atlantic Ocean which lay beyond the Pillars of Hercules. To this day the name has the same significance and differentiates the great open water masses from the seas, gulfs and straits, that lie around their borders. The Atlantic, the Pacific and the Indian, are the three great oceans of the world. The first, the grave of the mythical Atlantis; the second, named 'El Mar Pacifico', by Magellan, so calm were his first weeks thereon; the third called after the great country that bounds it on the north. In addition, the waters that surround the North Pole and those that lie along the coasts of the Antarctic continent, are termed the Arctic and Great Southern Oceans respectively.

Around the borders of these oceans lie the enclosed seas cut off by narrow straits, such as the Mediterranean, the Baltic Sea and the Persian Gulf; the fringing or partially enclosed seas, separated from the oceans by islands or peninsulas, such as the Bering Sea and the English Channel; gulfs and bays, such as the Gulf of Aden, Gulf of Maine and Bay of Bengal; and straits such as the Straits of Gibraltar and of Dover.

Over two-thirds of the earth's surface is covered thus by the oceans and their adjacent waters, the actual proportion of the water to the land masses being about 2·4 to 1. These water masses are not distributed evenly over the surface of the earth, only 43% lying in the northern hemisphere as opposed to 57% in the southern. The ratio of water to land also varies in the different hemispheres and it is possible to divide the earth into a water hemisphere whose centre lies a little south-east of New Zealand and a land hemisphere with a centre near the mouth of the Loire in France. Even in the land hemisphere the water area exceeds that of the land by a small amount, while in the water hemisphere only one-tenth is dry land.

Along the coasts of the great continents the water is comparatively shallow and a shelf is formed, either by erosion of the land through the ceaseless battering of the waves against its shores, or by the seaward extension of deposits of mud and silt brought down from the interior of the continents by great rivers, or by the

gradual submergence of the land itself. Thus there is a plateau, the continental
shelf (below) from which the dry land emerges above the water level. This area
of shallow water, extending down to a depth of about 200 m, varies considerably
in width. It is widest in those regions where there has been a gradual submergence of land, such as in the North Sea into which also are carried mud and
silt from the many rivers on the surrounding land. It is narrowest where there
has probably been an upheaval of land and where there is an absence of rivers to
extend, in a seaward direction, with their deposits these shallows already formed
by erosion through wave action. Such regions are on the western coast of North
Africa, and the Californian coasts of America. The effect of rivers in widening

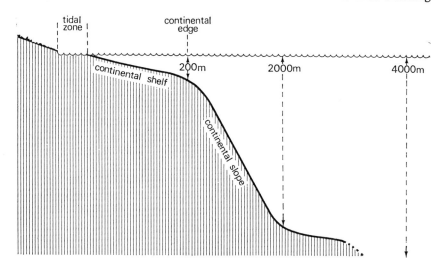

this shelf can be very clearly seen on the north coast of Egypt where it stretches
out for over 64 km in the neighbourhood of the Nile mouths, while 160 km west
of Alexandria the 200-m line is reached within 8 km.

From the outer edge of the continental shelf starts the continental slope
which is generally taken as stretching down to about 2000 m. This slope is comparatively steep and may be said to constitute the sides of the ocean basins. Its
upper limit, the 200-m line and outer edge of the continental shelf, is sometimes
known as the continental edge.

Formerly it was thought that beyond a depth of about 2000 m the slope of
the sea floor became almost imperceptible and was in most places not unlike a
vast and slightly undulating plain; that although it stretched down hundreds of
metres deeper, its extent was so great that in most places we should be unable to
appreciate any gradient whatever. From about 3600 m downwards this very
gradually shelving ocean bed was known as the abyss or abyssal plain.

It has long been known that the whole bed of the ocean is not absolutely flat
but presents considerable variation in depth over large areas. There is, for
instance, stretching north and south through the whole Atlantic Ocean a con-

tinuous ridge between 1800 and 3600 m in depth, surrounded on either side by water down to 8000 m deep (Plate 1). From this ridge rise the oceanic islands of the Azores, the Saint Paul Rocks, Ascension and Tristan da Cunha. It continues round South Africa and into the Indian Ocean where it divides, one branch running up to the Red Sea and the other going south of Australia into the Pacific. It has a deep rift valley all along its crest, and is volcanic in nature as recently proved by the eruption of Tristan da Cunha. It is now known that this ridge results from the gradual drifting apart of the continents.

The introduction of echo sounding made it possible to obtain many observations for every single sounding made with a lead. As a result our conception of the ocean bed became materially altered. Although over large areas the vast

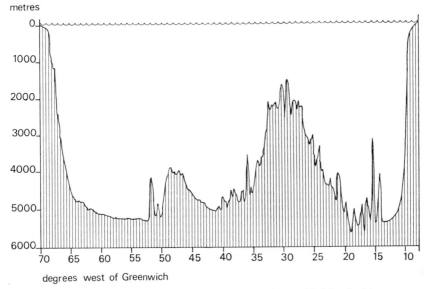

Cross-section of the North Atlantic Ocean to show mid-Atlantic ridge

masses of silt and bottom deposits may tend to smooth out the surface, it is now known that in many regions there are mountain ranges, peaks and cliffs just as we find on land. Most striking, for instance, is the discovery of canyons, such as those 1200 m deep off New York. Of special interest are the 'sea mounts' in the Pacific Ocean mentioned on p. 120.

The regions in which the bottom lies below a depth of 5500 m were known as 'deeps'. The greatest depth yet recorded is about 11,000 m in the vicinity of the Mariana Islands in the western Pacific. This enormous depth is hard to visualize. The reader can perhaps best realize it if he imagines that the highest mountain in the world, Mount Everest, be sunk upon this spot, when its loftiest peak would be fully covered and lie over a mile beneath the ocean surface.

Deeps occur in long, narrow areas known as 'trenches', which tend to run parallel with coastal mountain ranges. They are associated with regions in which earthquakes are prevalent. Such trenches are the Mariana Trench, the Philippine Trench, and the Japan Trench; but the greatest is the Tonga-Kermadec Trench, 2575 km long. In the past each deep was given a name, e.g. 'Murray Deep' and 'Valdivia Deep', after well-known oceanographers and research vessels. They are most numerous in the Pacific.

Taken as a whole the depth of the oceans is very great, for more than half of the ocean floor lies between 3600 and 5500 m, while well over three-quarters is deeper than 1800 m. One realizes how comparatively trivial in extent is the shallow water that lies around our coasts when it is known that the average depth of the oceans and their adjacent seas is over 3·5 km.

Ancient Beliefs

The extent of the oceans was not known to the ancients and their ideas of what lay beyond the small world they knew were probably mostly surmise and myth handed down from generation to generation. The Chaldeans imagined that the earth, which floated upon the eternal waters, was surrounded by a ditch in which a river perpetually flowed. The Egyptians likewise conceived ever-flowing around their world a river on the surface of which floated a boat carrying the sun. But neither of these civilizations can be said to have been maritime, and it was the Phœnicians who were the great navigators of ancient history. Their knowledge of the extent of the sea must have been very considerable, since they ventured often through the Pillars of Hercules, visiting the coasts of Europe and the British Isles. From their accounts of floating seaweed it has been thought that they may have been drifted by easterly winds as far west across the Atlantic as the Sargasso Sea.

But few of the Phœnician records have been preserved and we have no maps showing what they imagined the extent of the oceans to be. It seems that the Phœnicians whose business lay on the sea were unwilling to part with their knowledge and kept many of their trade routes secret. Any knowledge handed on from them to the Greeks was small and even that was vague.

In the days of Homer it was thought that the world was flat and that surrounding the Mediterranean was the land of the few countries they knew, but outside the land flowed an ever-running river which they called the ocean. In the sixth century B.C. Pythagoras considered that the earth was a sphere and in the fifth century Herodotus recorded further advances in knowledge. He, however, considered that the south and west of the then known world were bounded by the ocean, but could say nothing of the north and east. By the third century the idea of parallels of longitude and latitude had been introduced by Eratosthenes, and seventy years later Hipparchus instituted a method of map projection. This ancient era in map-making culminated in the description of the world by Ptolemy in A.D. 150, who imagined that the Atlantic Ocean and Indian Ocean were great enclosed seas, and that if one sailed into the Atlantic from the westernmost point of land the countries of the east would soon be reached.

Thus we see that the world of these ancient civilizations consisted of the lands to which they had access on foot and a few countries separated only by short

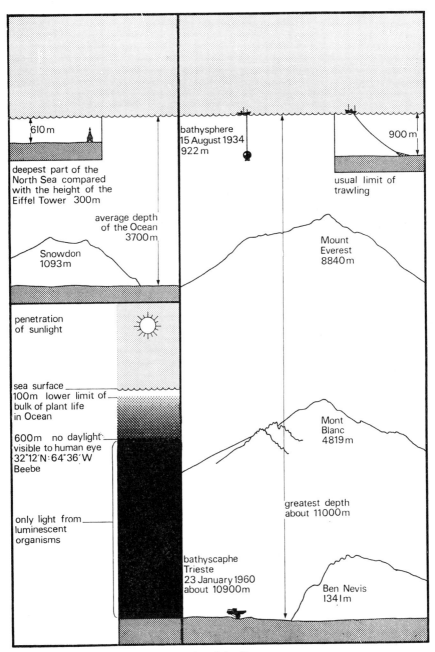

610 m

deepest part of the
North Sea compared
with the height of the
Eiffel Tower 300m

bathysphere
15 August 1934
922 m

900 m

usual limit of
trawling

average depth
of the Ocean
3700 m

Snowdon
1093m

Mount
Everest
8840m

penetration
of sunlight

sea surface
100m lower limit of
bulk of plant life
in Ocean

600m no daylight
visible to human eye
32°12'N:64°36'W
Beebe

Mont
Blanc
4819m

only light from
luminescent
organisms

greatest depth
about 11000m

bathyscaphe
Trieste
23 January 1960
about 10900m

Ben Nevis
1341m

Depths of the Ocean
1000 metres = 3282 feet = 547 fathoms

distances of sea, and the two oceans, the Atlantic and Indian, that bathed the shores of the known land. They had no knowledge of the Pacific.

From the fourth century A.D. onwards, when civilization suffered at the hands of barbarian invaders, knowledge of the geography of the world received a great setback. Fantastic suggestions were made as to the shape of the world, and Isidore of Seville in the seventh century originated the 'wheel maps' in which the world was represented as surrounded by a circle of water as had been thought in Homeric times. In addition the world was cut up into three portions by waters connected with this circle of ocean, an eastern portion, Asia, and two western portions, Europe in the north and Africa in the south. For many years little advance was made save for information obtained by Norsemen in the north and

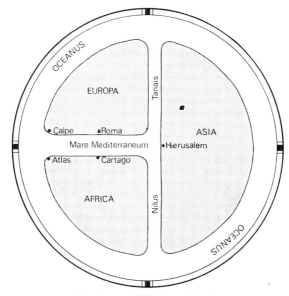

Wheel map of Isidore of Seville

voyages of the Arabs in the east, and in the fifteenth century the maps of Ptolemy reappeared. But the knowledge of the oceans of the world was soon to receive a great impetus, for from 1420 until his death in 1460, Prince Henry of Portugal, known as the 'Navigator', did all he could to encourage maritime research and exploration. During his lifetime much of the coast of western Africa and the eastern parts of the Atlantic were explored and charted, and his enthusiasm may be said to have prepared the way for the great voyages of exploration that were to follow. In 1486 Bartholomew Diaz rounded the Cape of Good Hope and thus opened up the connection between the Atlantic and Indian Oceans. On 12 October 1492 Christopher Columbus set foot on the island known by the natives as Guanahani, after having crossed the Atlantic Ocean. This island, which he named San Salvador, is generally thought to be that now known as Watling Island. For many years after this, expeditions sailed to the continent of America, but it was still thought that this continent was joined to the most eastern lands then known, as was considered to be the case in Ptolemy's time. In 1497 Vasco da Gama, rounding the Cape of Good Hope, reached India for the first time by sea. But the final link in the chain of knowledge of the great oceans was forged in September 1513, when Vasco Nunez de Balbao laid eyes upon the Pacific Ocean, a mass of glittering waters that dispelled for ever the idea that India and America were joined by land; and in 1520 Ferdinand Magellan sailed through the straits that bear his name into the 'Great South Sea' that he called the Pacific and, although he himself never lived to see the day, one of his vessels at last reached Spain in 1522, having circumnavigated the globe.

History of Oceanography

Let us pause to consider the true meaning of oceanography. It is one of the most composite of sciences, involving the many branches of knowledge that pertain to a study of the life and conditions of the environment in the oceans and seas. The study of the ocean currents, the tides, the temperatures, and the saltiness and other chemical properties, is approached through pure physics and chemistry. The charting of the ocean margins and the mapping of the relief of the ocean beds are matters for physical geographers and geologists. Intimately bound up with the great ocean currents and the tides are meteorological and astronomical phenomena. The above studies constitute hydrography, although this term is more usually applied to the survey and charting only of those features which have a direct bearing upon navigation. There remains the study of the life in the sea, marine biology, which involves zoology, botany, bacteriology, and includes the study of animal and plant development, of the abundance and interactions of life, and many other problems; to these must be added the application to the fishing industry of all knowledge of life and conditions of life in the sea, fishery research. This composite whole makes up the science of oceanography, which can be summed up as the study of the world beneath the surface of the sea.

It is evident, therefore, that up to 1522 when Magellan sailed into the Pacific Ocean the true science of oceanography had not begun; all these exploratory voyages had but touched the margin of the geographical side. Soundings were recorded also in ancient times, but as yet there was no systematic attempt to map in relief the ocean bed; in fact it is doubtful whether, owing to the backward state of other sciences, it would then have been possible. It is only in comparatively recent years that it has been made rapid and simple by the introduction of echo sounding.

The progress of oceanography, therefore, must march side by side with the advance of the sciences of which it is composed.

The first true oceanographical expedition sailed when Captain James Cook started on his voyage of discovery in 1768 in the *Endeavour*. On this expedition both temperature observations and deep sea soundings were made, and the ship's company included an astronomer, the noted biologists Banks and Solander, and several artists.

From this time onwards until about 1860 many exploratory voyages were

B

undertaken on which true oceanographical work was carried out, such as the expedition to the Antarctic in 1839 in the *Erebus* and *Terror* under Sir James Ross. At this period also knowledge of marine zoology and botany grew quickly and new facts were being discovered by naturalists who went out in naval surveying ships. In 1831 Charles Darwin sailed in H.M.S. *Beagle*; in 1846 H.M.S. *Rattlesnake* took with her Thomas Huxley, and in 1860 H.M.S. *Bulldog* went cruising with G. C. Wallich on board. Little was, however, known of the life on the deep ocean beds, and it was generally believed that the conditions found there would completely prohibit the existence of living animals. Occasionally organisms had been brought up attached to deep sea sounding-leads, but there was the possibility that these had become entangled while the rope and lead were hauled through the upper water layers.

More substantial evidence that life really existed at the great depths came

H.M.S. *Challenger*

when submarine cables were invented, and the salving of broken cables showed growth of marine organisms upon them.

In 1868, H.M.S. *Lightning*, followed later by H.M.S. *Porcupine*, with naturalists on board, proceeded to settle the question once and for all by deep-sea dredging. We can imagine the suspense of the little group of men on board when the dredge was at last nearing the surface; for deep-sea dredging is a lengthy business, the dredge taking many hours to be lowered and hauled through the enormous depths. They were not disappointed; living animals were present at the greatest depths; and an added excitement arose, for every haul brought up strange and beautiful organisms that man had never yet set eyes on.

The time was now deemed ready for a great expedition to explore all the oceans of the world, to study animal and plant life and the chemical and physical conditions under which they existed. Accordingly in 1872 Her Majesty's Government commissioned H.M.S. *Challenger*, under command of Captain Nares, carrying on board some of the most noted men of science of the day, headed by Sir Wyville Thomson, Professor of Natural History in the University of Edinburgh. This ship sailed the oceans for three years, travelling in that time 111,040 km (69,000 miles). Soundings were made the world over, temperatures noted, and samples of water taken from all depths; samples of the sea floor were obtained; dredgings were made in the great abyss, and nets towed in the water layers between the surface and the bottom.

The information obtained gave work to a large body of specialists and resulted in the famous 'Challenger' reports, which may be said to form the solid base upon which the superstructure of the science of oceanography has since been built. These reports were edited by Sir John Murray who took part in the expedition.

The results placed the science on a sure footing, and later expeditions went out with a good idea of the conditions to be expected, so that plans could be laid for special problems of interest that had to be tackled. Better instruments were invented and, most important of all, wire came into general use, replacing the old bulky ropes, Alexander Agassiz being the first to use it on the voyages of the U.S. ship *Blake*, from 1877 to 1880.

Expeditions sailed from many countries, and since the time of the *Challenger* some of the more important have been the Deep-Sea expedition of the Germans in the *Valdivia*, and the voyages of the *National* and *Deutschland* from the same country; the cruises of the Norwegian vessel, *Michael Sars*, under Professor Hjort and Sir John Murray of *Challenger* fame.

More recent cruises were those of the German *Meteor* and the Danish *Dana*, while oceanographical history was made by the Antarctic commissions in Captain Scott's *Discovery*, followed by R.R.S. *Discovery II* and *William Scoresby*. After the second world war oceanographical research was greatly developed owing to the need to increase the world's food supply from the sea and the military requirements for knowledge of the oceans. Expeditions by the Swedish *Albatross*, the Danish *Galathea*, and the Russian *Vityaz* have been followed by the almost continuous work of many nations. An inventory of the world research fleet in 1969 produced by the Food and Agriculture Organization of the United Nations numbered over a thousand vessels registered in seventy countries.

In the year 1901 was formed what is known as the 'International Council for the Exploration of the Sea'. This Council is supported by the governments of European countries. Its objective is the conservation and improvement of the northern fisheries, for these great fishing areas are truly international. Each country undertakes to maintain research ships and to study definite problems in the various areas of the sea allotted to it.

Since the second world war a number of similar organizations have been set up. Some of these cover prescribed areas of the oceans, such as North West Atlantic, or the North and the South Pacific. Others are special commissions for research on whales and seals and certain species of fish such as tuna and salmon, to decide what regulations might best conserve their populations. A number of these commissions were set up under the auspices of the Food and Agriculture Organization of the United Nations. More generalized marine research is now coordinated by a Scientific Committee for Ocean Research, under the International

Council of Scientific Unions, and by an Inter-governmental Oceanographic Commission.

International collaboration on a somewhat different basis was instituted in 1958. This was known as the International Geophysical Year, during which a number of nations made concentrated observations simultaneously on physical oceanography in certain sea areas. One of these was a study of the recently discovered Cromwell Current, a sub-surface equatorial current flowing eastwards in the Pacific. As a result of the success of this co-ordinated programme of research it was decided to make a chemical, physical and biological survey of the Indian Ocean, about which much less was known than about the other oceans. This took place mainly in the years 1962 to 1964 when thirty-eight ships from twenty nations made cruises on the so-called International Indian Ocean Expedition. One of the immediate fruits of this venture was the discovery of a valuable sardine fishery off the Arabian coast.

Marine Laboratories

The foundations of the science of oceanography were laid, as we have seen, by the work of the great marine expeditions. But the work that such expeditions can do is necessarily limited; they are indispensable for studies of the fauna and flora of the open sea, for examining the nature and inhabitants of the ocean bed, for studying ocean currents and the chemical and physical properties of the sea; but they are not suited, clearly enough, for examinations of the structure, and experiments into the functioning of the animals and plants, or for long continued investigations into the seasonal variations in the constitution of the sea water and its microscopical population, work which, as we shall see later, is of the very greatest importance. For these investigations it is essential that we should have laboratories on the sea coast where marine animals and plants can easily be obtained, where they can be kept in aquaria with running or circulating sea water as near natural conditions as possible, and where regular samples of sea water can be obtained throughout the year and the variations in its chemical and physical constitution and its contained life accurately determined by qualified chemists and biologists.

The need for such laboratories resulted in the foundation of a series of marine stations in the majority of the civilized countries, the work of which has been of the highest importance and promises, in the future, to be even more important. Expeditions remain necessary—we have still a very great deal to learn about the great oceans—but in the future they will extend and supplement the work of the shore stations rather than, as in the past, be an end in themselves.

Among the first and certainly the most famous of these stations was the one founded at Naples in 1872 by a German zoologist named Anton Dohrn. From the Italian government he obtained a grant of land situated between the town and the sea, where, with money from the German government and from scientific societies in different parts of the world, but very largely from his own private fortune, he built the famous Stazione Zoologica (Plate 2). Many times extended, this consists of flat-roofed buildings of white stone surrounded by evergreen oaks, palms, and cacti, and commanding from its upper windows a panoramic view of the Bay of Naples. All tourists to Naples will know it, for on the ground floor is situated the far-famed Aquarium, one of the recognized 'sights' of the city where as nowhere else the visitor can see something of the wonderful Mediterranean fauna of the Bay of Naples. The water which circulates through the aquarium tanks, and also the research laboratories on the upper floors, is pumped up from the Bay, and allowed to stand in huge storage tanks until the sediment has settled to the bottom when the clear water above is drawn off.

Scientists of all nationalities have worked at the Stazione Zoologica since its foundation, 'tables' being rented by the year by scientific institutions in different countries who have then the right to nominate workers.

Even more commanding in position and architecture is the combined museum, laboratory, and aquarium founded at Monaco by Prince Albert I, one of the most distinguished oceanographers of his day. The museum contains a unique collection covering all types of marine life and every form of apparatus used in the investigation of the sea, together with a model of the laboratory used by the Prince and his scientific staff during the oceanographic cruises of his ocean-going yacht the *Princess Alice*. This laboratory has recently taken on vigorous new life with the appointment as director of Captain J. Y. Cousteau, the pioneer of aqualung diving.

In this country the largest marine station is the Plymouth Laboratory of the Marine Biological Association of the United Kingdom (Plate 2), one of the leading institutions of its kind in the world. The Association was founded in 1884 and its laboratory, built in 1888, is continually being extended until it now provides facilities for many different branches of science, the permanent staff consisting of a director, eleven zoologists, three botanists, two chemists, two physicists, three physiologists and two biochemists. Facilities are also available for scientists coming from all parts of the world so that a great volume of work is annually produced. Much of this is concerned with seasonal changes in sea water and its contained population, work which can be done only by a resident staff. In all, some sixty workers can be accommodated in laboratories equipped with experimental tanks through which sea water circulates. There is a very fine library and also a public aquarium. For the collection of specimens and water samples and for making surveys there is a specially designed research vessel capable of working in deep water off the continental shelf and also smaller vessels for day-to-day work nearer inshore.

The Scottish Marine Biological Association has recently moved into a large new laboratory at Dunstaffnage just north of Oban (Plate 2). It formerly operated a marine station at Millport in the Clyde Sea area which originated in 1885, largely through the activities of Sir John Murray who had been a member of the *Challenger* expedition. The new laboratory has a permanent staff consisting of a director and some sixteen scientists and has appropriate research vessels. The Association was also responsible over a period of years for the Oceanographic Laboratory at Edinburgh which is concerned with the continuous plankton recorder invented by Sir Alister Hardy. This laboratory has recently been taken over by the Natural Environment Research Council and now forms a part of the Institute of Marine Environmental Research which is run by that Council. Other marine stations, situated at Port Erin on the Isle of Man, at Cullercoats in Northumberland, at Menai Bridge and at Robin Hood's Bay, are attached to the Universities of Liverpool, Newcastle, Wales and Leeds respectively. Fishery research is carried on in government laboratories at Lowestoft and Aberdeen. The Ministry of Agriculture, Fisheries and Food also has laboratories at Burnham-on-Crouch and at Conway which are concerned with shellfish, the former being also engaged in pollution studies. Many most important contributions to marine science in all its aspects come from the National Institute of Oceanography situated at Wormley in Surrey. Most of its work is physical but it also took over the staff and equipment of the *Discovery* expedition. While primarily concerned with problems of physical oceanography, it thus remains responsible for whaling research and various aspects of marine biological research in oceanic waters.

Only a few of the many marine stations which dot European coasts can be mentioned. The first were established in France, at Concarneau in 1859 and at Arcachon in 1863, but there are now many others, the most important, respectively, on Atlantic and Mediterranean shores, being at Roscoff and Banyuls. Despite her small population, Norway has produced much valuable oceanographical research due in part to the overwhelming importance of fisheries in her economy. The principal marine stations are at Bergen, Trondhjem and in the Oslo fjord at Drobak. Owing to the great depth of the fjords, these stations are ideally situated for the study of deep-water life. Sweden possesses a delightfully sited station at Kristineberg on the Gullmars Fjord in the Kattegat, while in Denmark the laboratory at Elsinore has recently been extended. Germany has now rebuilt the station she operated before the last war on Heligoland. The Dutch have replaced the laboratory at Den Helder just outside the Zuider Zee dyke by an enlarged, well-equipped building on the Island of Texel.

Outside Europe the most important marine laboratories are in the United States and Japan. In the former, the Marine Biological Laboratory at Woods Hole in Massachusetts (Plate 3) is the largest institution of its kind in the world. It is concerned with teaching and research on an immense scale. Close by is the Oceanographic Institution which has a large permanent staff and runs many research vessels; it is more concerned with physical than with biological oceanography. A third marine laboratory is primarily concerned with economic fisheries. Other major centres of marine research on the Atlantic coast are at Beaufort, North Carolina and at Miami. On the Pacific coast there are three important laboratories. The northernmost, at Friday Harbour at the mouth of Puget Sound, is run by the University of Washington, Seattle, which also has Departments of Fisheries and Oceanography at Seattle. In the central Californian coast is the Hopkins Marine Station at Pacific Grove on the Monterey Peninsula run by Stanford University, and in the extreme south is the vast Scripps Institution of Oceanography at La Jolla, near San Diego, which is a part of the great University of California. This possesses a positive fleet of research vessels but, like the Oceanographic Institute at Woods Hole, is more concerned with physical rather than with biological problems. In Canada there are laboratories concerned with fisheries problems at Nanaimo on Vancouver Island and at St. Andrews in New Brunswick.

No country exploits the products of the sea and also cultivates them to the same extent as Japan, which takes food from the sea over a wider area than any other nation. She has numerous marine stations concerned with every aspect of marine life and also many research vessels. This is also true of Russia which has laboratories scattered over its widely spread coast-lines in the Arctic, the Black Sea and in the North Pacific. She has the finest fleet of research vessels in the world. But almost every maritime nation now possesses at least one marine laboratory, the numbers of which grow year by year.

Life in the Sea

The sea is far richer in different forms of life than the land or fresh water, many groups of animals being exclusively marine. Since many of the latter may be quite unknown to readers without special knowledge of biology, it seems advisable, before proceeding with the description of the different zones of life in the sea and the characteristics of their inhabitants, to give in this introductory chapter a short summary of the various groups of marine animals and plants.

ANIMALS

Distinguished from all other animals, in that they consist of a single 'cell', are the PROTOZOA. They are extremely common in the sea; some are minute scraps of living matter, such as the marine amœbæ, unprotected and moving about on the sea bottom by a kind of flowing motion; others possess elaborate little shells of limy matter—these are the Foraminifera—or of silica constituting radiolarians, both occurring in countless numbers in the surface waters. Others again, the ciliates, though without shells, are of more complicated structure, being covered with fine ciliary hairs by the beating of which they move and draw in food. As widespread near the surface as the radiolarians and Foraminifera are the dinoflagellates which have two whip-like processes for locomotion, may or may not be covered with skeletal plates, and often contain green colouring matter so that it is uncertain whether they are animals or plants.

The sponges or PORIFERA are animals of so simple a structure that they give the impression of being little more than aggregations of protozoans. They are always attached and have a supporting skeleton, of horny matter in the sponge of commerce, but in the majority of cases consisting of immense numbers of tiny spicules of many beautiful shapes and formed in some cases of limy material, but more often of silica.

The CŒLENTERATA include a large number of relatively simple animals, their bodies essentially bag-shaped, containing only one cavity which combines the functions of the stomach cavity and the general body cavity of the remaining, more highly organized, animals. They are all built on a circular plan, i.e. there are not two symmetrical sides. They are all aquatic and mostly marine and are universally distributed. They include many common animals and may be said to be of two general types, attached, like the sea anemones, and freely swimming, like the jellyfish. Many of those attached are not solitary like the anemones but consist of many united individuals. Of such are the little hydroids which have branching stems dotted with many little polyps, like flowers, each one resembling a minute anemone, which are united to one another by a common canal which traverses the stem. The framework is usually of a thin, transparent, horny material, though in a few cases it is of limestone, the result being a stony coral. Some of these are not attached, have no skeleton, and float on the surface, an example being the Portuguese man-o'-war. Allied more closely to the anemones are the soft corals or alcyonarians, consisting of many individuals with a common skeleton of thick horny substance or of tiny spicules or other material, and also the true corals (Scleractinia) with massive limy skeletons.

The worms are divided into many groups with little resemblance other than in shape. There are the little flatworms or TURBELLARIA, seldom more than 2·5 cm long, very flat and almost transparent, to which are allied many parasitic worms which, though commonly found in marine animals, need not concern us here. There are also the NEMERTINA, soft-bodied worms with a long proboscis which they can protrude or draw in at will, and without that division of the body into a series of transverse segments which is so striking a feature of the most highly organized worms. These are known as the ANNELIDA and include the common earthworm (in which the segments we have just spoken of are very easily seen) and the bristle-worms (polychætes) which are almost exclusively marine, being everywhere abundant and constituting the great majority of the marine worms. Besides segmentation they are characterized by the presence of long bristles which project from the sides of the body. Some wander about freely and are called errant worms, but others live always in tubes of lime, sand or parchment-like material which they make for themselves, enlarging them as they grow and being in many cases able to make new ones if they are destroyed. Closely related are the leeches of which there are marine species which suck the blood of fish, and rather less so the sipunculids which have a protrusible proboscis and a tough leathery body without any sign of segmentation. They are all marine. One of the commonest constituents of the drifting life of the sea is the little arrow worm (*Sagitta*) which belongs to a small group called the CHÆTOGNATHA, which has no connection with other worms.

A new phylum of excessively thin worms, in tubes varying in thickness from fine threads to straws, has recently been identified. The animals, known as POGONOPHORA, live in bottom deposits. They have a fairly elaborate internal structure but, surprisingly, without a gut. They must absorb dissolved organic matter through the body wall.

Quite closely related to the bristle-worms are the POLYZOA or sea mats, minute creatures which always live in colonies, some of which are small and branching like the hydroids, and others large and with strong limy skeletons which give them the appearance of delicate corals. Some species grow as encrustations. In spite of their minute size the individual animals are very complex.

The ECHINODERMATA are a diverse group, exclusively marine, which include starfish and sea urchins. Like the cœlenterates, they are built on a radial plan though, owing to their mode of life, some have altered this and become bilaterally symmetrical. They all have limy internal skeletons, some consisting of continuous plates forming a compact shell, as in the sea urchins, and others of isolated spicules or tubercles embedded in a leathery skin, as in the sea cucumbers. They are in most instances either attached or very slowly moving animals, the mechanism of locomotion being usually provided by the peculiar 'tube-feet', tiny tubes which are worked by water power supplied by a series of canals within the body. There are five distinct groups, the starfish (Asteroidea) with a flat central disc to which are attached a number of arms, usually five though it may be many more; the brittlestars (Ophiuroidea) not unlike the starfish but with the disc sharply divided from the arms which always number five though they may divide and subdivide considerably, and without the groove which always runs along the underside of the arms in the starfish; the sea urchins (Echinoidea) which are all globular, heart-shaped or disc-shaped, with a firm skeleton or shell covered with spines and usually with definite rows of tube-feet; the sea cucumbers, sea gherkins or trepang (Holothuroidea) with sausage-shaped bodies along which run five rows of tube-feet; and finally the feather-stars and sea lilies (Crinoidea) sometimes attached and with a central disc with attachments below and above a series of five, often branching, arms which collect finely divided food particles from the water.

The ARTHROPODA include the largest number of species of any group in the animal kingdom. Like the annelid worms they have segmented bodies, but attached to all, or many, of the segments are the jointed limbs which give the group their name. Of the four great subdivisions, three—the insect, spider, and centipede classes—are almost as exclusively composed of land animals, as the fourth, called the Crustacea, are marine, one of the few examples of the latter found on land being the common woodlice. There is no space to go into the many subdivisions of the Crustacea, so all-embracing a group that it includes the tiny water fleas, the minute copepods which drift about in countless millions in the surface waters, the barnacles which cover rocks and the bottoms of ships, the sand-hoppers of the shore, the ghost-shrimps, the true shrimps and prawns, the lobsters and crawfishes, the hermit crabs and the many kinds of true crabs. The greatest difference between the simpler and more complex types is that in the latter the limbs, instead of being very much alike from one end of the body to the other, have become specialized into a number of feeding limbs near the mouth, then a group of walking legs, and finally a series of swimming legs under the tail region. Allied to them are the zoologically mysterious sea spiders or PYCNOGONIDA, resembling the true spiders in little beyond the possession of four pairs of long legs.

The MOLLUSCA form another great group. The Gastropoda or univalved shellfish, such as the limpets, periwinkles and snails, the great majority of which are marine (the chief exceptions are terrestrial snails and slugs), have usually a shell, always of one piece, though both land and sea slugs have either no shell at all or else one greatly reduced and often internal. Most of them live on the shore or on the sea bottom but a few, without or with greatly reduced shells, swim about near the surface, the sea butterflies being the commonest. The bivalve molluscs (Bivalvia) have a shell composed of two equal or almost equal valves with a uniting ligament which forces the two valves apart except when they are drawn together by the powerful adductor muscles. They are all either marine or fresh-water animals, well-known examples being mussels and oysters. The third great division of the Mollusca includes octopods, squid and cuttlefishes, and is called the Cephalopoda. Its members are all marine and are very highly organized animals with a head having a pair of complex eyes and surrounded with a series of arms possessing suckers and sometimes hooks. Some, such as the octopus, have no shell; others, the cuttlefish is an example, have a broad 'bone' within the back while a few, like the tropical pearly nautilus, have a large coiled and partitioned shell in which they occupy the last chamber.

Resembling the bivalves in the possession of a shell consisting of two valves are the BRACHIOPODA common only in deep waters in Northern seas. The two halves of the shell, however, are always dissimilar, while the internal structure of the animal is totally unlike that of the bivalve molluscs. The TUNI-CATA are exclusively marine and comprise many animals of very different appear-ance. There are the common sea squirts of shores and shallow seas, either solitary creatures like little gelatinous or leathery bags fastened on to rocks, or else great numbers of smaller individuals embedded in a common gelatinous mass, solitary and colonial tunicates respectively. Each individual has two openings which, when the animals are squeezed, squirt out a stream of water, hence their common name. Other varieties of tunicates, called salps and appendicularians, form an important part of the drifting life of somewhat warmer seas such as the Mediterranean. They are transparent animals, which may be solitary but are often fastened together in long chains.

In their early stages the tunicates resemble little tadpoles with a backbone which later disappears but the possession of which may mean that they are really somewhat lowly relations of the VERTEBRATES—distinguished from the INVERTE-BRATES, which we have hitherto been considering, by the possession of a back-bone. The simplest animals to show certain vertebrate characteristics are the lancelets, like little white eels which burrow in the sand, while, rather simpler than the true fish, are the CYCLOSTOMATA, of which the lamprey is our chief example. They have a larger number of gill openings than the true fish, have round mouths without jaws, and have no paired fins on the sides of the body.

The fishes are too well known to need description. They are divided into two principal groups, one, called the elasmobranchs, having a relatively soft, cartilaginous skeleton and with the gill openings separate—to mention two of their chief characteristics—and including the dogfish, sharks and skates, and the other, known as the teleosts or bony fish, with the cartilage replaced by bone, and the separate gill openings covered over with a flap known as the operculum.

All animals higher in the scale of evolution than the fish are air breathers but a number have returned to the sea and spend their lives in it, some always near or on the surface but others capable of diving to considerable depths though always compelled to return to the surface from time to time to obtain air. Among the REPTILIA are the water snakes or HYDROPHIIDÆ which have keeled bodies and tails flattened like those of fish to aid them in swimming, and the turtles which are aquatic tortoises with the limbs flattened to form swimming paddles. In the MAMMALIA there are the many kinds of whales, dolphins and porpoises which constitute the Cetacea and spend their entire lives in the sea, an existence for which they are just as well equipped as the fish themselves, and also the strange sea cows (Sirenia), sluggish, harmless beasts which usually live in shallow water near the mouths of rivers in tropical regions and frequently sit on their tails with only their heads out of water—the origin, probably, of all the stories about mermaids! Finally there is a group of the Carnivora (which include the lions, tigers, and bears) the members of which, almost exclusively marine except in the breeding season, are known as the Pinnipedia and include the seals, sea-lions, and walruses. In the north Pacific there are sea otters formerly hunted until thought to be extinct but now protected and slowly increasing.

PLANTS

The sea is far poorer in types of plant life than the earth. The flowering plants have very few marine representatives, the best known being the eelgrasses (*Zostera, Posidonia,* etc.), the former widespread in sheltered regions round British coasts. The great majority of marine plants are Algæ, of simpler structure than the flowering plants and reproducing themselves by 'spores' instead of by seeds produced by flowers. We may divide them into two types, fixed and drifting. The former are the seaweeds and may be further divided into four groups each of which has a characteristic colour as well as structure. There are (1) blue-green algæ (Cyanophyceæ), (2) the green algæ (Chlorophyceæ), (3) the brown algæ (Phæophyceæ) and (4) the red algæ (Rhodophyceæ); these are found from highwater mark downward into deep water in the order named; this is discussed in more detail later. The first-named are of slight importance—they are minute plants which sometimes form a slimy film over rocks—but many of the others are of considerable size, the brown algæ including not only the strange floating sargassum weed which gives its name to the Sargasso Sea, but also the largest known plant in *Macrocystis pyrifera* with fronds 183 m long, abundant in colder seas especially in both north and south Pacific. Green, brown and red algæ are common on all rocky shores. Since plants demand a certain minimum of light, all trace of weed disappears from the bed of the ocean below a certain depth. The deepest water from which weeds have been taken with any certainty that they were actually growing on the bottom appears to be about 120 m, but normally they are not abundant except in shallow water.

The drifting plant life, apart from the large sargassum weed, consists mainly of minute single-celled plants called diatoms which have tiny silica cases, and also the peridinians which, as stated above, can be considered either plants or animals. Both occur in untold billions and are of fundamental importance in the economy of marine life.

Chapter 2

THE SEA SHORE

That narrow strip between the high- and low-water marks of spring tides which we call the sea shore is the haunt of a rich and varied collection of plants and animals and has, on account of its unique position at the junction of sea and land, an interest altogether out of proportion to its area. Many books have been written about the sea shore and its inhabitants, for the subject is a big one and teeming with interest, but in this volume with its wider scope we can devote only one chapter to it and endeavour to give some general idea of the many fascinating problems it presents. Those who are sufficiently interested are advised to seek further information in the books recommended at the end of this volume.

The extent of shore uncovered at low tide naturally depends on the sharpness of slope, and this is the result of a variety of factors, the nature of the land, its configuration and the action of the tides, currents and rivers, being the most important. Anyone who has lived near the sea knows how in some places the out-going tide uncovers great areas of mud or sand, while in others, with a quickly descending shingle beach, only a comparatively small area is uncovered at the lowest spring tides. It has been estimated that the total area between tide-marks in Great Britain and Ireland amounts to some 279,000 ha. We can distinguish between three types of shore, formed of rock, sand or mud, though these may be mixed to a greater or less extent. The waves are the greatest influence at work moulding the shore, breaking with fury against the land, washing away loose material or, with the aid of pebbles and stones which they dash against it, gradually eating into a hard, rocky coast, forming a flat 'abrasion platform' at the base often of towering cliffs. Here we find a typical rocky shore. In other regions, on the other hand, the action of powerful cross currents deposits great banks of sand, formed by the breaking down of rocks. At the mouths of rivers or in sheltered creeks and gulleys there are mud flats where the sediment brought from land is deposited. The action of ice and of weathering in general also assists in the wearing away of the land and the formation of the shore, while the influence of plants and animals is not to be neglected. The former often help to bind together sand and mud, and convert them into firm, dry land; encrusting animals, such as barnacles and mussels, may help to protect rocks, but usually the action of animals is destructive, notably that of the various rock borers.

A feature of the greatest importance is the variability of conditions on the shore. Not only does the sea cover and uncover it twice daily, but the ranges of temperature are far greater, both yearly and often daily, than in any other region occupied by marine organisms. An animal in moderately cool water at high tide may be left stranded at low tide in a small rock pool where the temperature rises to great heights under the influence of a hot summer sun. The shallow water near the shore is both warmer in summer and cooler in winter (when ice may form on the shore) than the deeper water farther out. The constitution of the sea water is also apt to be variable especially near the mouths of rivers where much fresh water is mixed with it; also in the height of summer when, in enclosed areas, evaporation may make the sea more than usually salty. The influence of light is considerable, greater than in any other region except the surface waters of the open sea. This has an immediate effect on the distribution of plant life, for plants can only exist where there is light which is necessary for the photosynthetic action of their green pigment by means of which the carbonic acid gas in the atmosphere, or in solution, is combined with water to form starch. With the addition of salts obtained from the soil or from solution this is further elaborated to form proteins. The influence of this flora of seaweeds on the life of the sea shore is of great importance.

The population of the sea shore if it is to withstand these very variable conditions must be extremely hardy—how hardy we realize when we discover that very slight changes in the temperature or salinity of the water will kill animals used to the uniform conditions of the open ocean. Shore animals must be adapted in many different ways for protection, food collection, and reproduction, to mention only three most important aspects. Yet in spite of these difficulties the population of the sea shore is one of the densest and most varied on the surface of the earth. So dense, indeed, that the most striking features of shore life are the perpetual struggle for existence, the constant scramble for food in which the strongest or more subtle are the conquerors, the numerous devices for ensuring the continuance of the race, the never ceasing pursuit by the more powerful of the weaker and smaller, the latter often surviving to the extent to which they are able to disguise or hide themselves.

We can distinguish very definite associations of animals living on the shore. By this we mean that in different sets of conditions we habitually find the same types of animals, perhaps varying in species from place to place. These animals may be of many different kinds—some of them worms, others starfish, crustaceans, molluscs and so on—but they are all adapted for life under those particular conditions; one group may be plant feeders and live on weed, another may be attached and live fastened to rocks, another burrows into mud or sand, or they may be carnivores which prey upon the animals composing one of these communities to which they thereby become attached. The most intimate form of association is that of parasite and host (as we call the animal which 'entertains' the parasite), or the more equal type of intimate union known as symbiosis, some account of which is given in Chapter 9. Lastly, animals may always live together, one perhaps upon the other, a condition called commensalism or mutualism, because the two feed in common and also frequently assist one another. Examples of this are given later in this chapter.

There is, as shown by T. A. Stephenson, a universal pattern of zonation on rocky shores outside those scoured by polar ice. Species of littorine snails always occur highest, to above high-tide mark; below comes a zone dominated by

acorn barnacles but much influenced by the amount of weed and with a great
wealth of other animals. The level of extreme low water may be fringed by dense
growths of large brown seaweeds. All levels are influenced by the degree of
exposure, the various zones being denoted more by their populations than by
their actual height.

On temperate rocky shores we can distinguish these zones best by the different
levels at which the principal seaweeds grow. Of the different coloured weeds, the
green ones generally grow in pools near high-water mark or even above it where
sea water only occasionally penetrates and the water is largely fresh; the brown
weeds are especially common between tide-marks, as no one who has ever walked
on a rocky shore can fail to have realized, while below them come the red weeds.
Between tide-marks these are usually found at the bottom of the deeper rock
pools or when the spring tide uncovers an exceptionally large area, but are commonest in shallow water off a rocky shore.

Above high-water mark, except at the highest spring tides, there is an area of
mixed salt and fresh water—brackish is the term used to describe it—which, if
the ground be marshy, is usually characterized by the presence of the little salt-
wort, a tiny plant (not a seaweed) which has become adapted to these conditions;
if the shore is rocky and there are pools in this region they are filled with the
green weed, *Enteromorpha*, consisting of long, tubular fronds. In the summer the
pools are apt to dry up, the weed is killed and only a line of white marks this
Enteromorpha belt. About the region of high-water mark there is a zone varying
in breadth from about 1 to 5 m according to the slope of the shore, of the yellow
or olive coloured channelled wrack (*Pelvetia canaliculata*), distinguished from
the other brown weeds by its narrower fronds (Plate 3). From the base of this
belt to low-water mark the rocky shore is covered with a tangle of brown weeds
known generally as the fucoid zone, because most of the weeds are members of
the genus *Fucus*. The various kinds of *Fucus* are arranged in a very definite
series. Next to *Pelvetia* comes the broad fronded *Fucus spiralis*, followed by a
broader belt of *Ascophyllum nodosum* or knotted wrack, with exceptionally long
fronds bearing a single row of bulbous air bladders down the centre. This weed
is easily to be distinguished because on it invariably grows a small red weed
(*Polysiphonia fastigiata*). It is replaced in less sheltered waters by *Fucus vesi-
culosis* or bladder wrack (Plate 3), which has pairs of air bladders along the
fronds which, when dried, 'pop' sharply when pressed. Nearest low-water mark
is the commonest and most strongly growing of all these fucoids, *Fucus serratus*,
the notched wrack which, as its name tells us, has toothed serrations along the
edge of the fronds. These different belts of *Fucus* are not always present, depending on local conditions, the amount of fresh water, exposure, etc., while there are
several less common species which we have not mentioned, but generally speaking the above species will be found on any typical rocky shore. They are termin-
ated, quite sharply, at low-water mark by a broad belt composed of several species
of *Laminaria*, the largest of the brown weeds which have long very broad fronds
without a mid-rib, and are secured to the rocks by a massive holdfast. They are
never exposed except at the lowest spring rides.

A number of other weeds are common on the shore and can most conveniently be referred to now. The green weeds are especially widespread, the sea lettuce (*Ulva lactuca*) with broad, delicate fronds and living in pools usually above half-tide level, and *Cladophora rupestris* of a darker green and bushy compact growth which is common in pools everywhere. An easily recognized brown weed commonly exposed between tide-marks is *Himanthalia lorea*, which has peculiar cup-shaped attachments to the long narrow fronds, while of the red weeds, the dulse (*Rhodymenia palmata*) which has flat, irregular crimson fronds, and carragheen (*Chondrus crispus*) with thicker dark reddish-brown fronds which appear blue when seen in certain lights, are the commonest, occurring in the upper and lower regions of the shore respectively. The latter, known as Irish Moss, is eaten in certain parts of Ireland. Frequent in rock pools are the pink encrusting corallines (Plate 4), not at first easily recognized as seaweeds for they have limy skeletons like some of the encrusting animals.

Now let us consider some of the animals which live on rocky shores. They may be divided roughly into four categories, those which live exposed on the surface of stones, rocks, or weeds, those found under stones, those which live in holes and cracks in the rocks, and those inhabiting rock pools. Of the first group the most characteristic members are the sessile acorn barnacles (*Balanus* and *Chthamalus*) which cover the rocks with a carpet of little sharply pointed pyramids, especially near high-water mark where they form a definite Balanoid zone (Plate 6). Here too are many marine snails, of which the common limpet (*Patella*) and the periwinkles (*Littorina*) (Plate 7) and the top-shells (*Gibbula* and *Calliostoma*) are the commonest; they browse upon the weeds and corallines which cover the rocks. The limpets (Plate 7) may have definite homes on the rock face, to the surface of which their shells exactly fit so that by pulling down the shell, they can fix themselves when disturbed so firmly that very great force is needed to remove them; it has been estimated that limpets with a basal area of about 6·5 cm² (1 in²) need a pull of 32 kg (70 lb) to remove them. In all directions around them may be seen radiating paths cleared of weed, showing where the limpet has foraged for its food. There is a variety of periwinkles which live in different regions of the shore, one (*L. neritoides*) living very high, often where it is untouched by sea water for weeks together. The common winkle (*L. littorea*) lives lower on the shore always on rocks, while the smaller and less pointed species (*L. littoralis*), which varies in colour from black to white, is always found on the fronds of *Fucus*. Resembling the winkles but rather larger and with a thicker shell is the carnivorous dog-whelk (*Nucella*), which inhabits the upper half of the shore (Plate 11), preying upon the mussels or acorn barnacles. Other peculiar animals allied to the snails are the chitons of which the largest species are about 2·5 cm long and about half as wide; they are flattened, with the shell divided into eight plates, each slightly overlapping the one behind. This permits them to conform to the irregular rocky surfaces on which they browse; when pulled away they curl up like little armadillos. There is a numerous population growing on the weeds, notably many kinds of hydroids, sea mats and small snails.

Nearer low-water mark we find the rocks covered with sponges of which the crumb-of-bread sponge (*Halichondria*) is the commonest (Plate 7). It forms a dense mass which may be several centimetres thick and vary in colour from a pure white in shaded places to a yellow or green in more exposed situations. The deep crimson *Hymeniacidon* is often found with it. A common sponge which is attached by a stalk is the purse sponge (*Grantia compressa*), often found on the sides of rocks where there is no danger of it being dried up, and there are other kinds of sponges too numerous to describe. Growing on rocks or fucoid fronds is the tiny worm *Spirorbis*, which lives in a flat spiral shell like that of a snail and often occurs in great numbers. Larger worms with limy tubes of irregular shape (Serpulids) are frequently common on rocks. Generally covering rocks are the sea squirts, especially the compound forms—consisting of colonies of simple individuals—such as the golden-stars tunicate (*Botryllus*), which closely resembles a sheet of purple jelly dotted with tiny golden stars (Plate 4), though some simple sea squirts are found, especially the gooseberry (*Styelopsis gros-sularia*), a little reddish lump well described by its common name, and abundant near low-water mark. Other encrusting animals are the sea mats or polyzoa, which form, with the sponges, sea squirts, coralline weeds and other algæ, a dense carpet over the face of the rocks. Finally about low-water mark we may find bivalve molluscs such as the saddle-oyster (*Anomia*), the common mussel (*Mytilus edulis*) and some of the smaller scallops such as *Chlamys varia*. These are all attached by tough threads known as the byssus the formation of which is described on page 236. The scallop is secured by a limited tuft at the anterior end of the hinge and the mussels by many diverging threads which emerge from between the under side of the valves. The saddle-oysters lie on the right side with a massive byssus impregnated with lime issuing through a deep embayment in the under valve. This byssus is cemented firmly to the surface of the rock.

A fresh fauna is revealed when we begin to turn over loose stones. Near high-water mark there is a varied collection of small crustaceans and insects. Of the former the largest is the flat *Ligia*, sometimes nearly 4 cm long and closely resembling the common woodlouse; this is everywhere abundant and runs about, especially at dusk, usually just above high-water mark, for it is an animal by no means completely dependent on sea water, a slight moisture being apparently all that it needs. Associated with it is a number of sand-hoppers (*Talitrus, Orchestia, Gammarus*) which, when a stone or pile of dried weed is moved, go jumping away in all directions. Unlike *Ligia*, they are flattened not from above to below but from side to side. This lateral flattening is characteristic of amphipods whereas *Ligia* is an isopod. The insects include some beetles, but especially the little spring-tails. Lower on the shore, where the tide always reaches, there is a great variety of worms, such as nemertines, of which the most conspicuous is the long boot-lace worm (*Lineus*), usually in a complicated mass which is almost im-possible to disentangle without breaking, for it may be many metres long. There are also tiny flat-worms which move like semi-transparent films over the surface of stones or on the under side of the surface film in pools. Here we find great numbers of the more highly organized bristle-worms, examples of which are the bright green *Eulalia viridis*, which is common all over rocks, and the yellow *Cirratulus cirratus* which lives in patches of sand or mud beneath the rocks with only its filamentous tentacles to be seen. Still more worms are found nearer to low-water mark, the commonest perhaps the handsome *Nereis cultrifera*, of varied colours and 15 cm or more in length. It is frequently used as a bait, being known as rag-worm in many parts. A variety of little worms, broad and flattened with two rows of large scales on their backs, known as polynoids, are common everywhere.

Of the numerous crustaceans the largest and commonest is the common shore crab (*Carcinus*), examples of which scurry away from beneath almost every stone we examine. Common also is the velvet fiddler crab (*Portunus puber*), one of the largest of the swimming crabs, a very pugnacious beast, aptly called by the French fishermen 'Le Crabe Enragé', also a great variety of smaller crabs of varying shapes and habits, far too numerous to mention here. Some spider crabs are found, but they are commoner off shore and will be described in the next chapter. The curious little squat-lobster (*Galathea strigosa*) (Plate 12) may be mentioned, while occasionally left behind by the tide near low-water mark are small lobsters or edible crabs (Plates 12 and 41). Hermit crabs provide one of the quaintest and most characteristic members of the shore fauna. Unlike the other crabs, they have no shell covering the tail region but creep for protection within the empty spiral shells of molluscs, holding firmly on to the central column of the shell by means of claw-like appendages at the hind end of the body. They are able to shuffle about quite rapidly, carrying the shell, which they can, however, leave if they so wish, a procedure which becomes essential as they grow and have to forsake the old shell for another a size bigger. The peculiar little sea spiders, which have long legs and such thin bodies that the stomach for want of space has to penetrate the legs, are found under stones. Starfish are commonly encountered near low-water mark, especially the large red *Asterias rubens*, and, more difficult to see, the little cushion-star (*Asterina gibbosa*), of a dull grey or greenish colour which lives on the sides of rocks. Both of these have five arms but those of the latter are much shorter relatively, and are united for almost their entire length. Their relatives, the sea urchins, especially the small green *Echinus miliaris* which is a more typical shore form than the handsome *E. esculentus* (Plate 5), are also inhabitants of rocky shores. They are covered with spines— which in allied animals from warmer seas may be of much greater length, and sometimes thickness—and must be handled carefully.

Some shore fishes can always be found under stones. The most common is probably the butterfish (*Pholis gunnellus*), about 15 cm long, eel-like, flattened from side to side and with nine or more dark spots edged with yellow down the middle of the back. The smooth blenny (*Blennius pholis*) (Plate 4), can withstand long periods out of water and is common on the shore, as are the bullhead or father-lasher (*Taurulus bubalis*), with its head armed with four formidable spines, the five-bearded rockling (*Ciliata mustela*) to be recognized by the five barbels under its snout, and various kinds of sucker-fish (*Lepadogaster*), whose hinder (pelvic) fins are united to form a sucker with which the fish fastens itself to rocks.

In holes and cracks in rock live worms of various kinds, also a quaint crustacean called *Gnathia* and the little sea gherkins (*Cucumaria*) which can frequently only be removed by splitting the rock with a crowbar. Here, too, are rock-boring

c

bivalves, especially the common *Hiatella* (*Saxicava*), though the common piddock (*Pholas dactylus*) (p. 102) and some of its smaller allies are numerous in certain areas.

There is no more fascinating nor more beautiful spot on the shore than a typical rock pool. Its sides and bottom are usually covered with a many-coloured carpet of weeds, sponges, hydroids, sea mats and sea squirts, amongst which, like flowers, glow the rich colours of anemones (Plates 7 and 16). The commonest of these—by no means confined to the pools but common on the higher parts of shore—is the beadlet (*Actinia equina*), usually a deep red or brown, but some-times red with green spots (the strawberry variety) or, less frequently, a bright green. At the base of the tentacles is a row of blue spots. The largest of the anemones is the handsome dahlia (*Tealia*), with a warty column unlike the smooth one of *Actinia*, and of many colour patterns, some of them strikingly decorative. This also is not uncommon on the shore at the base of rocks, but usually nearer low-water mark. Especially common in the pools is the snakelocks anemone (*Anemonia sulcata*)—appropriately so named for it has long tentacles which wave with the motion of the water and, unlike those of the other anemones, are in-capable of contraction. We must pass by many of the other anemones, mention-ing only the beautiful little *Corynactis*, very common in some regions, no bigger than a pea and of many colour-varieties, pink, green and white ones being found. Other inhabitants of rock pools are hydroids, of which the handsome *Tubularia* is the finest, a variety of tube-dwelling worms of which the most con-spicuous is *Bispira*, with parchment-like tube and yellow crown of tentacles in the form of two spiral whorls united at the base, many snails especially the conspicuous sea slugs (Plate 12), such as *Aeolidia*, with its back covered with a 'fur' of soft grey projections, the sea lemon (*Archidoris*)—also very common on rocks everywhere—and a great number of smaller species. Crustaceans abound, especially prawns, both the large *Palaemon* (p. 121) and the little Æsop prawn (*Hippolyte*), so difficult to see because of its remarkable power of colour change.

A sandy shore has a very different population. Both seaweeds and encrusting animals are absent for there is no hard surface to which they can attach them-selves. Because of the need for protection when the tide is out, all sand dwellers are burrowers, only making contact with the surface when the tide is in. The predominant animals are bivalve molluscs which do not attach permanently like the mussel or the saddle-oyster, but burrow by means of the muscular foot which is protruded from between the shell valves. This dilates terminally by blood pressure to grip the sand; muscles attached to the shell are then contracted so pulling the bivalve spasmodically along.

Broadly speaking there are two types of these animals, one well exemplified by the common cockle (*Cerastoderma* (*Cardium*) *edule*) and the majority of sand, as distinct from mud, burrowing bivalves and including species of *Venus*, *Spisula* and of the razor-shells (*Ensis* and *Solen*). These are all suspension feeders, drawing in water through the one siphonal opening, sieving its contents through the gills (see page 230) and passing the water out through the other, upper, siphon. The other type of bivalves are deposit feeders; in them the siphons are completely separated; one gropes over the surface drawing in fine detritus, and water is then expelled through the shorter, exhalant siphon. The commonest such bivalve on sandy shores is the delicate and laterally very flattened *Tellina tenuis*. The related wedge shells (*Donax vittatus*), with a very beautiful and highly polished shell, are inhabitants of exposed sandy beaches.

Worms are numerous in sand. Some of these also are suspension feeders such as the terebellids which have a crown of ciliated tentacles and live in sandy tubes, the upper ends of which, fringed with fine filaments, project above the surface. The larger *Amphitrite* is an inhabitant of muddy sand, usually amongst rocks. In sand also are various carnivorous worms, such as *Nephthys*, which move actively within it. Where there is some mixture of mud, and so of organic

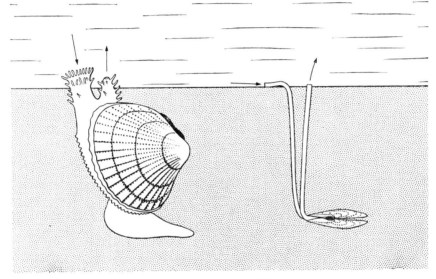

Left, a cockle, *Cerastoderma edule*, a superficial burrower and example of a suspension feeding bivalve with short, united siphons. Right, the thin tellin, *Tellina tenuis* with long separated and very mobile siphons, an instance of a deeply burrowing and deposit-feeding bivalve.

matter, the well-known castings of the lug-worm, *Arenicola*, may be common (Plate 10). This worm inhabits a U-shaped tube, throwing up castings at the hind end but with a characteristic depression in the sand over the head end. This surface sand containing edible organic debris is pulled down the head shaft shown in the figure and swallowed by the worm, which from time to time backs up the tail shaft to extrude the castings. And all the time, as indicated by the solid arrows, a current of water for respiration is drawn through the burrow from tail shaft to head shaft. Under ideal conditions of ample organic matter within the sand, populations of lug-worms may reach 32,800 per hectare. An animal easily mistaken for a worm is the reddish *Synapta*, found in sand along southern shores, which is a sand-burrowing sea cucumber. Allied to this—

in structure, not appearance—is the burrowing sea urchin (*Echinocardium cordatum*) (Plate 14), to be dug around low-tide mark. But it is commoner on sandy bottoms offshore.

It is on such sandy shores that the common shrimp (*Crangon crangon*) is frequently found although difficult to see owing to its sandy colour and its habit of covering itself with sand, leaving only the long feelers exposed. With it are a variety of other crustaceans, such as the ghost-shrimps (mysids), smaller and even more difficult to see. There are several kinds of crustaceans which construct burrows often of great depth so that considerable industry is needed in digging

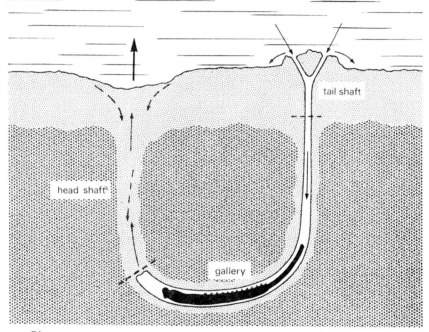

Diagrammatic representation of the burrow of the lug-worm, *Arenicola marina*, with the animal *in situ*. Explanation in text. (After Wells)

them. Several anemones habitually live buried in sand with only the mouth disc and the surrounding tentacles exposed. Where reefs of rock run out into the sand we frequently find large colonies of the peculiar reef-building worm *Sabellaria* (Plate 10), which forms a sandy tube not in, but above, the sand, and not singly but in great numbers altogether, so that large reefs of hardened sand are formed which, on examination, will be found to be honeycombed with the tubes of the worms which construct them. Various fishes are common in shallow water on sandy shores, being occasionally left behind in pools by the retreating tide; of such are young flat-fish of various kinds and a variety of sand-eels, Ammodytidae.

The change from the clean sand of an exposed beach to the mud flats which usually fringe an estuary is accompanied by a change in population. We now find animals that occupy a permanent instead of a temporary burrow; because of the amount of organic detritus food is often abundant, but for this very reason oxygen is usually low—used in oxidation of this organic material. Amongst bivalves the deep-burrowing gaper (*Mya arenaria*), the American soft shell clam (p. 237), may be dug from depths of up to 30 cm. Although the shell is large, 10 cm or over in length, it is exceeded by the siphon with which it maintains

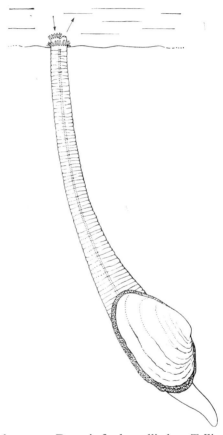

The sand gaper or soft shell clam, *Mya arenaria*, a deep burrowing bivalve with long fused siphons (i.e. a suspension feeder), the shell covered with dark horny periostracum, and with a reduced foot. (After Meyer and Mobius)

contact with the surface for food and oxygen. Deposit feeders allied to *Tellina* are the more rounded *Macoma balthica* and, higher up the estuary, the much larger and extremely flattened *Scrobicularia plana*. In both, the groping inhalant siphon can extend for remarkable distances. Burrowing worms occur, but especially those living in tubes such as the common peacock worm (*Sabella pavonina*), with the projecting rubbery tubes and widely spread ring of red and white plume-like tentacles, and also *Myxicola infundibulum*, with a thick and translucent gelatinous tube and a shorter more compact ring of tentacles. *Sabella* may be so common that at low water the tubes, which project for some

12·5 cm above the surface, form a miniature forest. The only anemone adapted for life in mud is the small burrowing species *Peachia hastata*. There are various crustaceans from the common shore crab to the immensely abundant little burrowing amphipod *Corophium* and the massive burrowing 'shrimps' *Upogebia* and *Callianassa*. All the last three occupy permanent burrows. The surface of the mud may be granular with the small shells of the little snail, *Hydrobia*. Over it also move the scavenging whelks, species of *Nassarius* (*Nassa*), with herbivorous periwinkles if there are stones. A variety of prawns and also sticklebacks and flounders penetrate far up rivers, being apparently indifferent to the change from salt to fresh water—a barrier which effectively prevents the great majority of shore animals from passing up estuaries.

We have dealt with the plants and animals of the sea shore very briefly, it is true, but in sufficient detail we hope, to give some idea of their variety and of the different types found under different conditions; it is now necessary to consider the many peculiarities of these shore animals and the particular devices they possess which enable them to exist and propagate their kind in the strange section of the earth's surface which they inhabit.

Methods of Attack and Defence

Of the first importance are means of attack and defence. Purely mechanical weapons are not particularly common and are best exemplified by the powerful pincer-like claws of the larger crustaceans, especially lobsters and crabs. In the former the two pincers are different; if a lobster be examined it will be found that one of the claws is larger with rounded, irregular teeth—clearly adapted for crushing—and the other more slender with numerous sharp teeth and is used for cutting. There is no regular arrangement of these claws which may occur on either side; they are immensely strong and, in the case of the shore crab, have been found capable of supporting a weight equal to about thirty times that of the body, whereas a man's right hand when clenched is unable to support a weight equivalent to that of his own body! Poison is largely used, especially by the anemones and hydroids and all their allies, which possess batteries of stinging cells, the action of which is described on page 137. All over the surface of the shell of sea urchins are little clawed spines, of various patterns but usually with three teeth which can open and snap together. These are used to clean the shell, removing any fragment of waste matter, but are also for protection. If an enemy attacks an urchin the long spines on that side turn away, exposing the smaller clawed spines beneath, which snap at any part of the enemy touching them and also produce a poison which enters into the wound.

More universal are methods of passive defence. The thick shell of the crustaceans and of the univalve and bivalve molluscs are examples. The last-named are usually safe so long as the shell can be firmly closed. The two muscles concerned with this—the adductor muscles of the shell—are extremely powerful, though very variably so in different cases, those of the cockle for example having less than a quarter the power of those of *Venus*. How such bivalves may be successfully attacked and eaten by whelks and starfish we shall see in Chapter 9. The hermit crabs have found an excellent means of protection by using empty snail shells. Then there are the various devices whereby an animal escapes attention by toning with the background. This may be done by having a colour similar to the surroundings—one particular set of surroundings or any surroundings, the animal in the latter case changing its colour to tone with the background. This capacity is common in crustaceans and flat-fish on the shore (see Chapter 8). The vivid colours of many sea slugs, which are unable, however, to change colour, sometimes tone with those of the rocks or weed on which they live. But such naked snails have poison glands in the skin and are avoided by experienced predators. Many spider crabs mask themselves by deliberately decking their shells with pieces of weed or sponge which continue to grow there. The dahlia anemone and the smaller sea urchin both cover themselves with stones and pieces of shell. In all these cases we must remember that the concealment is not only from foes

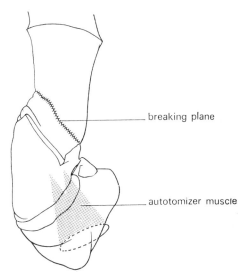

breaking plane

autotomizer muscle

Basal portion of the second walking leg of the lobster, *Homarus vulgaris,* showing mechanism of autotomy, i.e. severance of the limb at a special breaking plane by contraction of autotomizer muscles used solely for that purpose.

but also from the prey, i.e. for attack as well as defence. For protection alone are the limy, sandy or parchment-like tubes of the worms, while the borers find the interior of stone or wood safer than a more exposed habitat.

The replacement of lost parts of the body is a very familiar occurrence on the shore. Crustaceans have an almost unlimited capacity for growing new limbs. It is not unusual for them to be faced with the alternative of sacrificing one or more limbs or else losing their lives. They do not hesitate to do the former and the limb is cast off at a special line of weakness called the 'breaking plane', the fracture being caused by the deliberate contraction of the muscles at this point. Owing to the depth of the furrow, the wound is small and blood quickly coagulates and closes the opening. So the animal remains, with a stump in place of a limb, until the next time it moults, when the rudiments of the new limb force their way out quickly before the new shell has had time to harden. At each successive moult the process is carried a little further until the fresh limb is as large as those which were not injured. This habit of *deliberately* parting with limbs which are later regenerated is called 'autotomy' and is not confined to the crustaceans, being widespread amongst the starfish and their delicate allies, the brittlestars. Both of these part with their arms very readily and quickly grow new ones; they may lose all their arms and yet from the central disc there will grow out another complete set.

The related sea gherkins have the more unusual power of casting up their viscera when disturbed or attacked, then proceeding at their leisure to grow a new set. It may well be that, under normal conditions, the attacker is satisfied with the meal of soft entrails thus provided, and will not trouble their owner further. The worms have exceptional powers of regeneration and when cut in two accidentally—or by design in the laboratory—will grow new heads or tails with equal facility. Sponges have probably the most remarkable powers in this direction, for if a sponge be broken into tiny fragments which are then strained

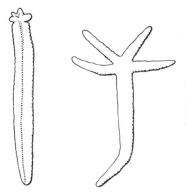

Two stages in the regeneration of new arms from a single arm of a starfish, *Asterias rubens,* retaining, however, adequate portions of the central disc. (Modified after Flattely and Walton).

through fine meshed silk, these isolated pieces will come together and unite to form a new individual!

We are leaving consideration of parasitism and of the intimate, mutually advantageous union of two animals or an animal and a plant, known as symbiosis, to Chapter 9, but we must say something here of that looser form of association called 'commensalism', which may be defined as an external partnership between two different animals usually for their mutual benefit. On the shore the most striking example is furnished by the association between hermit crabs and different kinds of anemones and sponges. One species of hermit (*Eupagurus prideauxi*) is always found with its body enfolded by an anemone (*Adamsia palliata*). This resembles a sausage-shaped bag pushed in at either end to form, at the upper end the stomach cavity, the mouth of which is surrounded by tentacles, and at the lower end the much larger cavity occupied by the soft body of the crab. A snail shell is always initially present and to this the crab is attached, the anemone extending the cavity so that the crab never needs to find a new shell. The common hermit (*E. bernhardus*), which lives in shells of all sizes up to those of the whelk, usually carries on the shell a large anemone (*Calliactis parasitica*), while in the upper whorls of the shell may be a worm called *Nereis fucata* (p. 154).

Yet a third, smaller hermit (*E. pubescens*) is often almost obscured by the relatively large masses of a sponge which grows on the shell. This definite association between the hermit and the anemones or sponge is clearly not accidental; in the former case the anemone probably helps to protect the hermit which, in turn, provides the anemone with scraps of food; in the latter case the sponge may provide protection by camouflage, and itself receive food.

Other examples of commensalism are provided by gall crabs which live in certain corals. The young female crab settles down between two branches of the coral which, as a result, broaden and finally unite above the crab, forming a gall within which the crab lives, feeding not on the coral, but on particles suspended in the water. The much smaller male remains outside the gall. The pea crab, which lives in bivalves and sea squirts is a further example. Another is furnished by a tropical crab (*Melia tessellata*), which carries anemones, deliberately removing them from the rocks, in each of its claws. It holds them fully expanded, pushing them forward when attacked and taking out the food they capture and transferring it to its own mouth. The large claws are used exclusively for carrying the anemones. The advantage to the crab is obvious while the anemone, in spite of losing so much of its food, perhaps also gains. By being moved about in this manner it has so many more opportunities of obtaining food.

Adaptations for Respiration

The problem of respiration—of obtaining that essential minimum of oxygen, without which no animal can exist—is often serious on the shore, where the inhabitants are part of the time in water and the remainder in air. We have not space here to discuss the different organs, or gills, used in different animals for obtaining oxygen, but some instances of shore animals which are able to respire both in air and in water will be of interest. This can be done, in a sense, by a variety of animals which are able to keep their gills moist for considerable periods while out of water, for example crustaceans, which have gills covered over by the edge of the shell—as in the larger forms like the lobster and crabs—or in the form of plates beneath the body. Of the different periwinkles, those which live highest on the shore and may be out of water for long periods, have developed the power of breathing air, the walls of the gill cavity becoming covered with fine blood vessels so that the blood is able to take up oxygen which is first dissolved in a superficial layer of moisture. There are crabs which live exclusively on land, some, such as the tropical robber crab, only going to the sea to breed, which have developed a true lung for breathing, while others which have retained their gills have to go down to the sea occasionally in order to moisten these. There are also species of tropical fish, notably *Periophthalmus*, the mud skipper, which live out of water; they have a lung-like extension of the gill chamber, while it is also reported that they respire through their tails, for they sit on land with only their tails in water! If held under water for any length of time they drown.

Locomotion and Migrations

As we have seen, many shore animals are attached to rocks or weed or else live in permanent tubes or burrows. The advantage of this mode of life is clear when we consider the perpetual beating of the waves on the shore as the tide rises and falls. Of equal advantage is the burrowing habit of many animals, bivalve molluscs and burrowing sea urchins, for example, which are also able to move about beneath the surface. Movement in shore animals takes many different forms. The larger crustaceans clamber over rocks and through gullies by means of their strong, hinged walking legs. The starfish move steadily over the surface by the concerted action of the many tiny 'tube-feet', each terminating in a small sucker. Double rows of these feet line the grooves which run down the centre of the underside of the arms. They are connected with a complicated system of canals containing water which can be forced out of or into the tube-feet at will, enabling

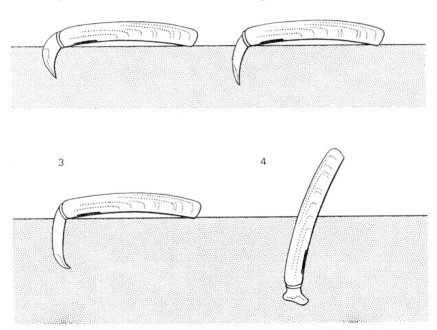

Razor-shell (species of *Ensis*) showing successive stages of burrowing into sand. Process described in text. (After Fraenkel)

them in turn to fix and relax their hold. A similar mode of movement is found in the sea urchins, which also employ their teeth for this purpose! A mussel may move up the side of a rock by fastening a byssus thread as far as possible above it and then pulling itself up by hauling on this. Some shore fish, such as the blennies, can crawl over rocks by means of their fins. The shore insects and the crustacean sand-hoppers move by jumping, suddenly straightening out the tail or hind end of the body. Periwinkles, whelks and all their allies really glide over the surface, waves of movement passing over the broad flat sole of the large fleshy foot. Anemones and flat-worms glide in a similar manner. Apart from the fish many animals swim, the lobsters by sudden movements of the tail which propel them quickly backwards through the water, while many worms are able to swim to a greater or less extent by undulatory movements of the long body, some, most highly specialized, having paddle-like flaps on the sides of the body. Mention will be made of swimming crabs and of swimming scallops in the next chapter. The peculiar movements of boring animals are discussed in Chapter 6. The burrowing bivalves work their way through the sand by the action of the muscular foot, the most specialized case being that of the razor-shell, which drives its pointed foot directly downwards, anchors it there by forcing in blood which makes the end swell out and then, by a sudden contraction, draws the shell down to the level of the foot, repeating the process as often as need be and burrowing so quickly that one has to drive a spade in very suddenly to capture it. Burrowing worms, crustaceans and sea urchins all have special devices which enable them to work their way through the sand or mud.

Many fishes migrate long distances, usually because their feeding and spawning grounds are far apart. There are also migrations on the sea shore, though on a much smaller scale. The movement of edible crabs has been studied, and it has been found that they move from near the shore into depths of 40 or 60 m about the beginning of autumn and remain there until February, when they begin to return towards the shore. The eggs are laid in the winter in deep water and remain attached to the body of the female, finally hatching out during the summer in the warmer water near the shore. Both temperature and food, therefore, play a part in determining the yearly travels of the crab and also of lobsters and prawns which, though we know less of their habits, appear to migrate in a similar manner. The shore fishes also move into deeper water in the winter, while many of the bigger sea slugs, such as the sea hare (*Aplysia*) and the large plume-bearer (*Pleurobranchus*) come on to the shore during the summer for spawning. Attention has already been drawn to the more localized feeding movements of limpets.

Spawning

Spawning is a matter of supreme importance and the special problems of shore life, especially of exposure both to violent water movements and to the air, have to be solved if the species is to be continued. Temperature is the major factor affecting the time of spawning and in the majority of (although not all) shore animals this takes place in spring or early summer when the water teems with microscopic plant life which forms ideal food for newly hatched young.

We may divide shore animals into three groups according to the mode of reproduction. There are those which discharge their reproductive products freely into the sea where they float helplessly until swimming organs are developed; those which deposit adhesive spawn on to stones, seaweed or empty shells; and those where eggs and developing young remain within or on the body of the parent. In the first division are sea urchins, starfish, brittlestars, many worms, barnacles and the bivalve molluscs. Since the chances of destruction are excessively great, the number of eggs shed is correspondingly large, a striking example of which is supplied by the American oyster (*Crassostrea virginica*), which may produce over 100,000,000 eggs at a spawning and may do so several times in the year! If a sea urchin or mussel be watched when spawning, it will be seen to discharge a cloudy fluid which, on microscopical examination, will be found to consist of incredible numbers of eggs or sperm. Yet so great are the dangers to which these unprotected young are exposed that the species does no more than hold its own!

If search be made among the rocks during the spring, many kinds of spawn will be found. Common on the underside of rocks are the egg capsules of the dog-whelk (Plate 11), which look like grains of corn, and are attached by short stalks, occurring in groups of fifty, a hundred or even more. Each consists of a tough case, containing a number of developing embryos, amongst which there is the keenest competition, the weaker being eaten by the stronger, so that finally only the one or two strongest emerge. The sea slugs provide the most diverse and ornamental spawn, covering the rocks with ribbons of jelly, often beautifully coloured, in which the developing eggs lie embedded. The common sea lemon (Plate 11) lays its eggs in a broad band of pure white, always arranged in a triple coil, some 38 cm long and about 2·5 cm wide, the margin being unusually wavy. Over half a million eggs may be laid in a single ribbon, for the young hatch out at a very early and unprotected stage. The spawn of other sea slugs show other peculiarities, some consisting of spirals with up to ten whorls, the edges of the ribbon being scalloped or the whole zig-zagged in its course, and usually white or yellow in colour though that of the common *Æolidia* is pink and ropy.

The eggs of the crustaceans are usually carried by the female—either in 'brood pouches' on the underside of the body or, in the more highly-organized crabs, prawns or lobsters, attached to the swimming legs—until the early stages of development have been passed. This may take several months, so that the young leave the parent, not as defenceless eggs or embryos but as actively swimming 'larvæ', a full account of which is given in Chapter 5. A crab or lobster which is 'in berry' is probably familiar to everyone, and shows this condition clearly. In the sea spiders the male carries the eggs, which are handed over to him in a bundle by the female as soon as they are laid.

The acorn barnacles, resembling molluscs rather than crustaceans, species of *Balanus* and *Chthamalus*, which cover immense areas of intertidal rock, are hermaphrodite but fertilize each other (Plate 11). The young develop within the shell cavity to be extruded as very obviously crustacean larvæ, nauplii with three pairs of limbs which temporarily dominate the plankton. After a series of moults, these change into the more elongated 'cypris' stage when they settle, attaching by way of cement glands in the antennae and suffering profound changes in structure and habit (see p. 152).

The varied manner in which closely related species may spawn is well shown by the habits of the different species of periwinkles. Strangely the two species that live highest and lowest on the shore, the small *Littorina neritoides*, which may often be above all but the highest tides, and the large and common *L. littorea* (the one collected for food), are both completely dependent on the sea into which they release their spawn. The very rounded *L. littoralis* which lives on the large fucoid weeds in the mid-shore lays its eggs in gelatinous masses on the surface of the weed. Finally the rough periwinkle (*L. saxatilis*), which lives on the upper half of the shore, is viviparous although individuals, probably con-stituting separate species, have been found which lay eggs.

Not all animals reproduce sexually. Some do so asexually, producing buds, dividing in two or breaking into fragments, each of which grows into a new individual. The anemones, although some kinds habitually reproduce themselves sexually, often divide (a process which is highly developed in the reef building corals, as described in Chapter 7) or break off small fragments near the base. The little syllid worms break up in a perfectly definite way into fragments a few segments long, each of which grows a new head at one end and a new tail at the other, and so develops into a full-sized worm. The hydroids have a complicated history. Many produce not eggs but little jellyfish, called medusæ, which swim about near the surface of the sea and, when fully developed, produce eggs (or sperm) which develop into the hydroids which produce the medusæ. This, in common with the production of swimming young by various shore animals, is of great importance in securing the distribution of animals which, because of their slowness of movement or fixed mode of life, would not otherwise be able to spread far.

Growth

Length of life in shore animals varies within the widest limits. Some animals, such as many sponges and sea squirts, are annuals or even go through several generations in a single summer, whilst on the other hand anemones may live to a great age; there are some in captivity known to be over sixty years old. After the early embryonic or, in the case of animals with swimming young quite unlike their parents, 'larval' stages, most animals assume the adult shape only changing in size as time passes. Growth is usually a steady process, varying in rate accord-ing to the time of the year, being generally slower in winter and quicker in summer when it is warmer and there is more food. In the crustaceans, however, growth takes place by a series of jumps. Unlike the molluscs, which increase the shell by adding to the free edges, or the sea urchins, which increase theirs by adding new material to plates already formed, the crustaceans are unable to add to the shell and have to cast it and form a new one. These moults take place at regular intervals, depending on the particular species and on the prevail-ing conditions, the entire shell being cast—not only that which protects the body but the covering of the limbs and the lining of the stomach and gills and of the hinder part of the gut! The animal withdraws itself, often with obvious difficulty, from its old shell and is then left soft and completely defenceless. Although the new shell has been formed under the protective cover of the old one it takes time for this to harden even though the materials needed have been stored in readiness. The new shell is always larger than the old one and the animal immediately begins to swell by the inward passage of water until the soft covering is stretched tight, in which state it proceeds to harden by deposition of lime. All such animals grow by a sudden increase immediately after the hard shell is cast.

Chapter 3
THE SEA BOTTOM

Bottom Deposits

Before we discuss the animals which inhabit the bed of the ocean at different depths and in different zones, we must first say a little about the nature of the bottom. In limited regions where there are great water movements this may be of scoured rock, but more usually, and invariably in deep water, it is covered with deposits, the nature of which formed a great part of the life-work of the famous Scottish oceanographer, Sir John Murray, who was one of the scientists on the *Challenger* expedition. It is to him that we are indebted for the basis of our knowledge on this subject, for, as a result of his exhaustive study of samples collected on that expedition and on later expeditions, he was able to classify the ocean deposits into a series of groups which later research has completely confirmed.

Broadly speaking, there are two types of bottom deposits. There are those formed by the settling of material washed down from the land by rivers or worn away by wave action, which Murray called terrigenous deposits and which are found in deep and shallow water near the land. The only regions where this type of deposit is found any great distance from land are opposite the mouths of the greatest rivers, such as the Amazon, which bring down vast quantities of material in suspension which the strong current of the river carries far out to sea. It might be thought that material in suspension would not sink readily in the sea, but this is not so because, as a result of the mixing of the fresh water with the sea, the fine particles become attracted electrically to one another and so form accumulations which readily sink just at the region of mixing of the fresh water and the salt.

At about 200 m the action of waves and of transporting currents ceases to be effective, and below this line, which is called the mud-line and corresponds roughly to the continental edge, the particles come to rest and from this region outwards the bed of the ocean is covered everywhere with mud although this is not necessarily at rest. Particularly within deep submarine canyons which in places cut through the margin of the continental shelf, great masses of mud continually cascade to greater depths. The nature of the terrigenous deposits depends on the nature of the land from which they are derived. Between tide marks and in shallow waters they consist of sand, gravel, shingle and boulders with mud in sheltered areas; rather further from the shore are gravel, sands, banks of living and dead shells, and, especially opposite estuaries, banks of mud.

If the land is volcanic this is reflected in the volcanic nature of the deposits, if of coral origin the deposits are formed of finely divided calcium carbonate while, off continents, they consist usually of quartz gravels and sands together with marls. The terrigenous deposits beyond the mud-line were divided by Murray into five categories—blue, red, green, volcanic and coral muds, the first three varying somewhat in their chemical content but having much in common, while the nature of the last two is clear.

In areas far from land the deposits on the sea bottom consist exclusively of material which has dropped down from the surface waters. This second type of marine deposit was termed pelagic by Murray (Plate 17), and is made up of volcanic dust which has fallen on to the surface of the water from the air, or of dead animals or plants which, as they die, rain down upon the bottom far below. The vast bulk of life in the open sea is composed not of large and active creatures, such as fish or whales, but of microscopic plants and animals which drift near the surface and make up the all-important plankton. It is their skeletons which form the great proportion of the deep-sea deposits, for their bodies soon disintegrate or are eaten by other animals. From their consistency Murray called these deep-sea deposits oozes, and distinguished between them according to the relative abundance or complete absence of the various kinds of skeletons.

Particularly around the poles, in the Southern Ocean especially, the surface waters contain vast numbers of the microscopic plants called diatoms (Plate 32), each individual enclosed in a delicate case of silica, and it is these minute plant skeletons which form the chief constituent of the deposits in these regions, hence called diatom ooze. Although there is an abundant flora in the temperate and tropic seas, there is a larger proportion of animals, those with calcareous shells being the commonest. Chief among these are the tiny Foraminifera, of which one called *Globigerina* is the most plentiful, and calcareous ooze, with its limy skeletons as the principal constituent, is extremely widespread, covering an estimated area of some 124,272,000 km², being especially abundant in the Atlantic. A third type of deposit which is really only a variety of the last is called pteropod ooze; it takes its name from the predominance in it of the limy skeletons of delicate swimming snails known as pteropods or 'sea butterflies' which are commonest in warmer seas where this type of ooze is exclusively found. It is always in shallower water than globigerina ooze and it is commonest near coral islands and on submerged elevations far from land.

Beneath a certain depth oozes with limy shells as their principal constituent are no longer found, all calcareous matter having been dissolved on its passage downward through the water. In certain areas, notably in the tropical regions of the Pacific and Indian Oceans, there are great numbers of delicate protozoans known as radiolarians in the surface waters. These have skeletons of silica, often of very beautiful lattice design, which form the most conspicuous part of the deep-sea ooze in these regions (Plate 17). Finally, at the greatest depth of all (below 5400 m) even the silica is dissolved and the deposits then consist exclusively of what Murray called 'red clay'. This is a true clay which has been formed by the prolonged action of the sea water on volcanic dust which is all that has survived the long journey from the surface to the bottom. Like the siliceous deposits it is present at all depths but is usually masked by the deposits of animal or plant origin. In this red clay occur spherules probably of meteoritic origin and also teeth of sharks and ear bones of whales, often belonging to species now extinct, both being almost indefinitely resistant to solution.

Life on the Sea Bottom

These bottom deposits do not have the overwhelming importance of soil on land because even the seaweeds do not root in them, usually attaching to rock by an extensive holdfast but always obtaining their nourishment from salts in solution in the sea. But they are nevertheless of great importance to the bottom-living animals, known collectively as the benthos to distinguish them from the largely surface-dwelling plankton and the actively swimming nekton of many depths. The great bulk of the benthos consists of invertebrates and these may be divided into those which either attach themselves to a hard surface, or else crawl freely over this, and those which burrow into soft deposits. These constitute the epifauna and the infauna respectively. The former consist of animals such as sponges, many cœlenterates from the minute little plant-like growths of hydroids to the massive sea trees (gorgonids) and the corals which form extensive reefs, also the little sea mats, the barnacles, some molluscs such as cemented oysters and mussels attached by their byssal threads, and the sea squirts. We have already encountered many of these between tide-marks on rocky shores. With these attached animals are many others which crawl over a rocky bottom, including crabs and other crustaceans with a wealth of worms, starfishes and other echinoderms and many kinds of marine snails. The much more abundant infauna—because so much more of the sea floor is covered with deposits—comprises a host of animals with the common capacity for burrowing. Except briefly, few live on the surface. They live in burrows or they plough their way through the bottom material. Bivalve molluscs are probably the commonest of such animals but many crustaceans and worms are highly adapted for burrowing together with some sea urchins and starfishes. Members of the infauna are confined to definite types of substrate, some to coarse gravel, some to sandy gravel, others to pure sand, to sandy mud or fine mud. The bottom deposits are sorted out in this manner owing to the action of the water, the more rapid its movement the fewer and coarser particles are dropped out of suspension; the slower the movement the finer the particles until in regions where there is little or no movement, often in depression and always below the mud-line, the finest particles drop to the bottom. This mud, it should be noted, contains the finely divided organic remains of animals and plants; it is therefore more than a medium in which the animals burrow. It is a major source of food, in part because of the bacteria and the protozoa with other minute animals that feed directly upon it.

The animals found in any grade of soft bottom (and also on rock) are collectively termed a community. They are associated because each is fitted for life in the same type of bottom, not because they depend on one another. The

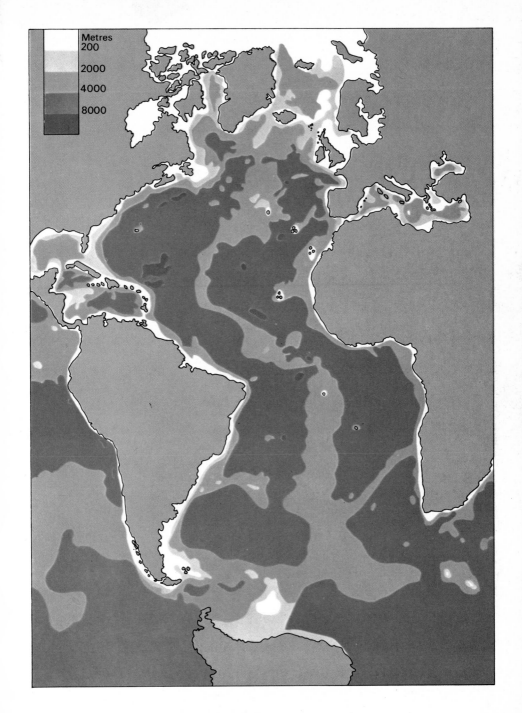

PLATE 1   Map showing depths of Atlantic Ocean
1000 m = 3282 feet = 547 fathoms

Laboratory of the
Marine Biological
Association at
Plymouth

Laboratory of the
Scottish Marine
Biological
Association at
Dunstaffnage, near
Oban, Argyllshire

PLATE 2

Marine Biological
Laboratory, Naples

(*above*) Marine
Biological
Laboratories,
Woods Hole, Mass.,
U.S.A.

Flat winkle,
Littorina littoralis,
on bladder wrack,
Fucus vesiculosus
(Photo: Heather
Angel)

PLATE 3

Channelled wrack,
*Pelvetia
canaliculata*, with
broader flat wrack,
Fucus spiralis
(Photo: Heather
Angel)

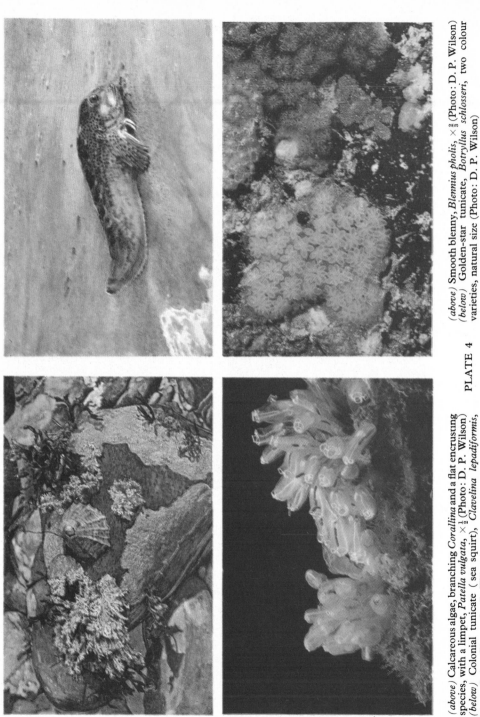

(*above*) Calcareous algae, branching *Corallina* and a flat encrusting species, with a limpet, *Patella vulgata*, × ½ (Photo: D. P. Wilson) (*below*) Colonial tunicate (sea squirt), *Clavelina lepadiformis*,

(*above*) Smooth blenny, *Blennius pholis*, × ⅔ (Photo: D. P. Wilson) (*below*) Golden-star tunicate, *Botryllus schlosseri*, two colour varieties, natural size (Photo: D. P. Wilson)

PLATE 4

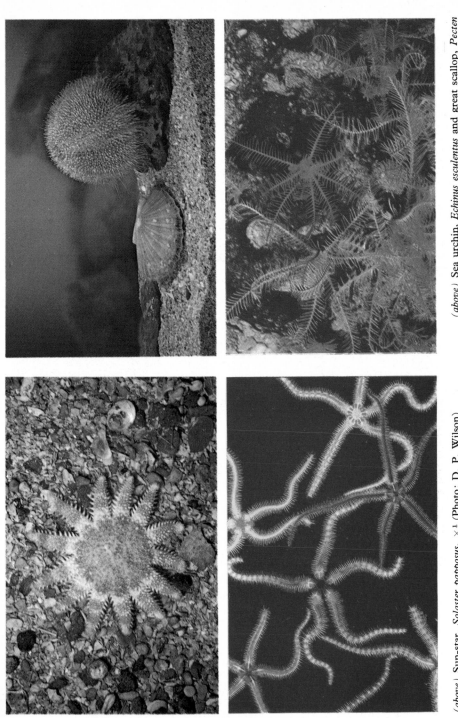

(above) Sea urchin, *Echinus esculentus* and great scallop, *Pecten maximus*, ×⅓ (Photo: D. P. Wilson) (below) Feather-stars, *Antedon bifida*, ×½ (Photo: D. P. Wilson)

(above) Sun-star, *Solaster papposus*, ×⅓ (Photo: D. P. Wilson) (below) Common brittlestars, *Ophiothrix fragilis*, ×⅔ (Photo: D. P. Wilson)

PLATE 5

Acorn barnacles, *Balanus balanoides*, of three age groups with a limpet, *Patella vulgata*, in bottom corner, natural size (Photo: Heather Angel)

PLATE 6

Rock face with dog-whelks, *Nucella lapillus*, largely in depressions abrupt under margin of barnacle zone indicated by arrow (Photo: Heather Angel)

Beadlet
anemones,
Actinia equina,
and limpets,
Patella vulgata,
of varying sizes
with winkles
exposed at low
tide, $\times\frac{1}{3}$ (Photo:
Heather Angel)

PLATE 7

Common
periwinkles,
*Littorina
littorea*, $\times\frac{1}{2}$
(Photo: Heather
Angel)

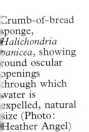

Crumb-of-bread
sponge,
*Halichondria
panicea*, showing
round oscular
openings
through which
water is
expelled, natural
size (Photo:
Heather Angel)

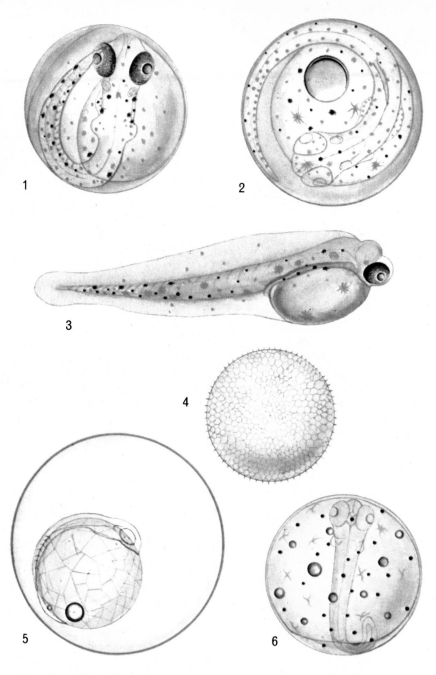

PLATE 8   FISH EGGS AND LARVA, × ca. 33

1 Whiting, *Merlangius merlangus*. 2 Gurnard, *Eutrigla gurnardus*. 3 Whiting, *Merlangius merlangus*. 4 Dragonet, *Callionymus lyra*. 5 Pilchard, *Sardina pilchardus*. 6 Weever, *Trachinus vipera*. (After H. O. Bull)

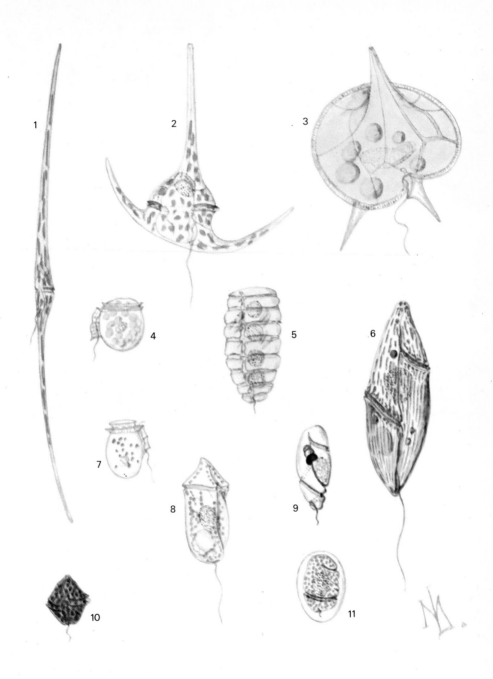

PLATE 9   DINOFLAGELLATES, × ca. 220

1 Ceratium fusus. 2 Ceratium tripos. 3 Peridinium depressum. 4 Dinophysis rotundatum. 5 Poly-krikos schwarzii. 6 Gyrodinium britannia. 7 Dinophysis acuminata. 8 Gymnodinium lebourae. 9 Pouchetia polyphemus. 10 Gonyaulax polyhedra. 11 Cochlodinium schuetti. (After M. V. Lebour)

PLATE 10

(above) Casts and hollows of the lug-worm, *Arenicola marina*, on rippled sand, $\times \frac{1}{2}$ (Photo: D. P. Wilson)

(below) Colonies of the reef-building or honeycomb worm, *Sabellaria alveolata*, natural size (Photo: D. P. Wilson)

Dog-whelks, *Nucella lapillus*, on rocks with recently deposited egg capsules; barnacles on left and below, $\times \frac{1}{2}$ (Photo: D. P. Wilson)

PLATE 11

Sea lemon, *Archidoris pseudoargus*, laying its gelatinous egg-ribbon in an aquarium, natural size (Photo: D. P. Wilson)

Acorn barnacles, *Balanus crenatus*, showing extended penis transferring sperm from one immovable individual to another, $\times 2$ (Photo: Heather Angel)

(above) Squat-lobster, Galathea strigosa, ×⅔ (Photo: D. P. Wilson) (below) Coral reef sea slug, Hexabranchus sanguineus, natural size (Photo: T. E. Thompson)

(above) Young edible crab, Cancer pagurus, ×⅔ (Photo: D. P. Wilson) (below) Common British sea slug, Aeolidia papillosa, natural size (Photo: T. E. Thompson)

PLATE 12

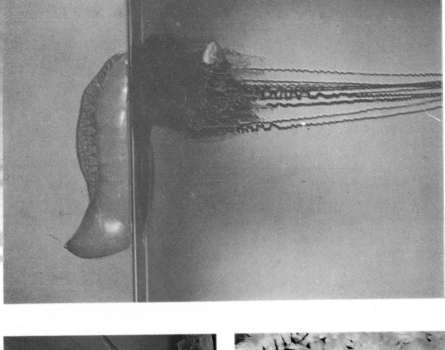

(*above*) Sea fan, *Eunicella verrucosa*, with calcareous polyzoan ross, *Lepralia*, and branching sponges, ×$\frac{1}{2}$ (Photo: D. P. Wilson) (*below*) Sea fan, *Eunicella verrucosa*, with polyps expanded, ×$1\frac{1}{2}$ (Photo: D. P. Wilson)

Portuguese man-o'-war, *Physalia physalis*, consuming a fish, ×$\frac{1}{3}$ (Photo: D. P. Wilson)

PLATE 13

PLATE 14

(below) Spider crab, *Inachus dorynchus*, partly covered with encrusting sponge that aids in concealment, natural size (Photo: D. P. Wilson)

(above) Polyzoan sea mat, *Electra pilosa*, with extended individuals, ×2 (Photo: Heather Angel)

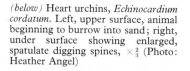

(below) Heart urchins, *Echinocardium cordatum*. Left, upper surface, animal beginning to burrow into sand; right, under surface showing enlarged, spatulate digging spines, ×⅔ (Photo: Heather Angel)

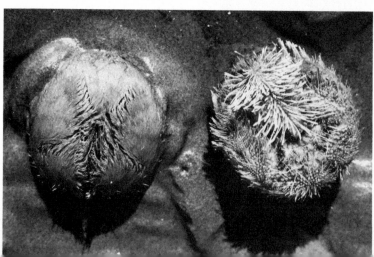

PLATE 15

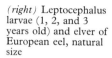

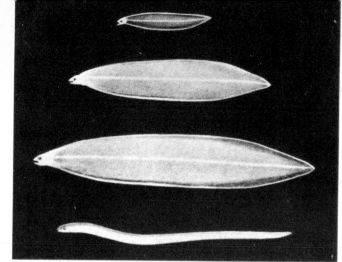

(right) Leptocephalus larvae (1, 2, and 3 years old) and elver of European eel, natural size

(below) Leptocephalus larva and elver of American eel, natural size

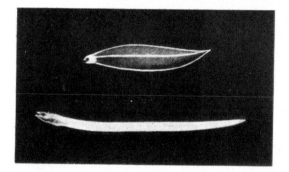

(below) Common oarweed, *Laminaria digitata*, exposed at low spring tide showing stalk or stipe and broad fronds

(*above*) Beadlet anemone, *Actinia equina*, feeding on a prawn, ×1 (Photo: Heather Angel) (*below*) Snakelocks anemone, *Anemonia sulcata*, ×⅓ (Photo: Heather Angel) (*above*) Plumose anemone, *Metridium senile*, ×1 (Photo: Heather Angel) (*below*) Dahlia anemone, *Tealia felina*, ×1 (Photo: Heather Angel)

PLATE 16

existence of these communities was established by the Danish marine biologist, C. G. J. Petersen, who was the first to sample the sea bottom quantitatively by taking localized bites out of it by means of a grab (Plate 34) instead of pulling a dredge indiscriminately over it. He named the communities he found after

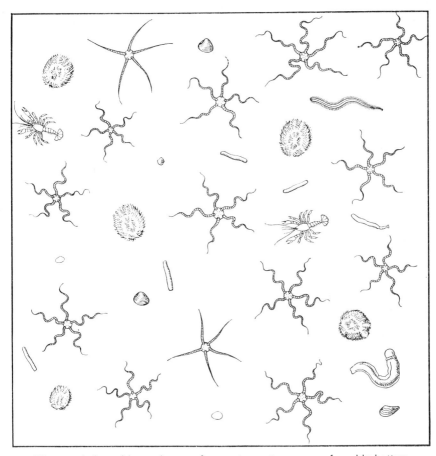

The population of invertebrates of a quarter-metre square of muddy bottom characterized by a particular abundance of burrowing sea urchins, *Brissopsis lyrifera* (dark, oval-shaped), and burrowing brittlestars, *Amphiura chiajei* (with undulating arms), constituting Petersen's *Brissopsis–chiajei* community. (From Moore, after Petersen)

particularly common or conspicuous members such as the mud community shown here which is characterized by the burrowing sea urchin (*Brissopsis lyrifera*) and the brittlestar (shown with much bent arms) (*Amphiura chiajei*), and is known as the '*Brissopsis-chiajei*' community.

The great majority of these bottom-living animals spend their early larval life as temporary members of the surface-dwelling plankton. They settle from this

D

on to a bottom which is likely, more often than not, to be unsuitable for adult life, but under these conditions they will delay change, or metamorphosis, into the adult form for some time. During this period they may well chance to drift on to the suitable bottom when at once they begin to change into the adult form and then assume the adult habit of life.

Generally speaking bottom-living animals are most numerous in shallow waters, as far as the edge of the continental shelf; then they diminish in numbers with increasing depth. But life exists in the very greater depths. The Danish *Galathea* expedition of 1950 to 1952 dredged animals (and also obtained bacteria) from the bottom of the Philippine Trench, 10,190 m below the surface.

Differences in fauna are chiefly the result of differences in depth, temperature and food, with salinity also in some enclosed seas like the Baltic. Food is controlled by factors affecting general productivity in any area which, as we shall see, is largely dependent on the extent to which the water is mixed. Temperature has a major effect in shallower waters and changes in this are primarily responsible for the great dissimilarities in the fauna and flora of tropical, temperate and polar regions—differences in horizontal distribution. The effect of temperature is most rigorous on spawning because animals may be able to live in water both too cold and too warm for spawning; but the species clearly cannot maintain itself except within the limited temperature range in which it can reproduce itself. Temperature and depth are the major factors controlling the vertical distribution of animals from the shore into deep water. Owing to the decrease in temperature, the horizontal differences become less and less as the water gets deeper, thus we find similar animals at great depths in the polar and equatorial regions because the temperature is the same, while animals found in shallow water in temperate seas, for example, may have relatives living in deeper water with a similar temperature in the tropics.

The shallow waters of the globe can be divided into different regions corresponding to changes in the horizontal distribution of animals, the result, as we have seen, of differences in temperature. These, it must be noted, do not follow parallels of latitude, but are greatly influenced by cold and hot currents, an example of which is provided by the western coasts of Europe bathed by the warm waters of the North Atlantic Drift and the frozen coasts of Labrador—in similar latitudes—which are washed by the cold Labrador current. We can distinguish between an Arctic region, including the water round the North Pole, and 'boreal' areas along the west coast of Europe and in the North Pacific; an Indo-Pacific region including the shores of India, East Africa, China, the East Indies and Australia; a West American region along the tropical shores of West America; an East American region running southward from Newfoundland; a West African region including both the Mediterranean and the Guinea coast; and finally an Antarctic region which includes the southern coasts of Australia, Africa, and South America from the Argentine on the east to Ecuador on the west. Within any one area the members of the marine fauna have much in common. The fact that a number of animals from the Arctic and Antarctic regions are closely akin has led to the development of the theory of 'bipolarity', which postulates that the animals inhabiting these cold seas are more closely related to one another than they are to the animals inhabiting the warm waters which stretch between them. It is thought that they maintain connection by way of the cold deeper waters of the tropics. It may equally be true that the animals from the two polar regions which resemble one another are both descended from the same deep-water animals of intermediate regions which found their way into shallower water at both ends of the globe.

The sea bed can be divided into a series of zones each occupied by a largely

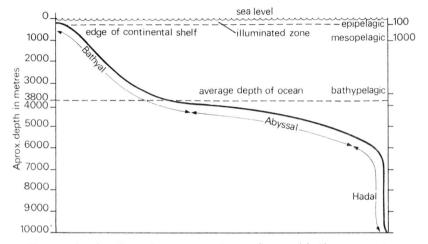

Diagram showing the major zones on the sea floor and in the open water.
(Modified after Hedgpeth)

characteristic assemblage of animals which in the deeper zones are little influenced by latitude, the temperature altering little. Below the Littoral, or intertidal, stretches the Sublittoral Zone usually subdivided into inner and outer areas, the former down to some 40 m, the latter extending to the edge of the continental shelf at around 200 m. The continental slope and the broad abyssal plain down to some 4000 m comprise the Bathyal Zone which is replaced below that depth by the Abyssal Zone. This was formerly regarded as the deepest zone but, following the work of the Danish *Galathea* deep-sea expedition, there is now good reason for introducing a Hadal Zone. This is below 6000 m and consists largely of the areas, or 'deeps', represented by the great ocean trenches.

Sublittoral Zone

In the previous chapter we discussed the life of the shore which constitutes the intertidal or Littoral Zone. There is, however, a sublittoral fringe to this, areas which are only exposed at low water of spring tides and where we can during those periods examine animals and plants which are truly members of the inner sublittoral. This fringe can be roughly subdivided into areas occupied by *Laminaria* or by *Zostera* or consisting of exposed rock, of sand or of mud.

In the first, characterized by veritable forests of tangled *Laminaria* (Plate 15), a large population lives attached to the weed, among the most common of which are sea mats, especially *Membranipora membranacea* (Plate 15) which forms a delicate latticed encrustation with typical rounded outlines where the colony is actively growing over the surface of the frond. Many hydroids are present, their branching stolons ramifying over the surface, the branches bearing the feeding polyps projecting freely into the water. A little limpet (*Patina pellucida*), distinguished by the blue stripes on its shell, is invariably found browsing on the bulbous holdfasts of the *Laminaria*. A great assemblage of animals find a home in this region, many of them so small as not to be discovered without great care; there are small sea slugs which feed on the hydroids and sponges, and many little mussels, worms, snails and starfish, some of which spend their entire lives here and others only the early stages. In the holdfasts of the weed are worms and

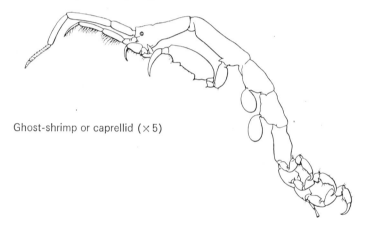

Ghost-shrimp or caprellid ($\times 5$)

one of the boring bivalves, *Hiatella*. Perhaps the most interesting of the many crustaceans are the grotesquely elongated Caprellids, or ghost-shrimps, which are especially adapted for climbing about on weeds and hydroids, being provided with grasping claws at either end of the body progressing rather like a looping caterpillar.

Zostera is an eelgrass, many kinds of which occur in different seas, which lives in sheltered pools and estuaries usually rooted in mud, for it is a flowering plant, not a seaweed, and has roots for absorbing nutriment from the soil. In some parts of the Kattegat and in the lagoon-like Limfjord in Denmark, it covers vast areas. The associated fauna is not very rich; little snails, species of *Rissoa* are very plentiful and this is also the haunt of the beautiful green *Haliclystus*, best defined as a little, squarish jellyfish which crawls about on the eelgrass. This is the home of fish such as sticklebacks and pipe-fishes, the latter, which have long, slender, brown or green bodies, being very difficult to see against the dead and living blades of eelgrass. They swim in an upright position by the vibration of the delicate dorsal fin, coming to rest in the same position and swaying with the motion of the water. Closely related to them are the quaint sea-horses. A number

of snails lay their egg capsules on eelgrass, on which, however, there are few encrusting animals. Cuttlefish may live here.

A large and characteristic fauna occurs on hard bottoms in the Sublittoral
Zone. There is an attached epifauna, notably of sponges, sea mats, hydroids, sea
squirts and corals, the latter including the little solitary cup coral (*Caryophyllia*),
one of the few northern examples of stony corals, and the 'soft coral' dead-
men's-fingers (*Alcyonium digitatum*) which consists of dead-white (sometimes
yellow) finger-like growths covered with tiny feeding polyps). Serpulid worms

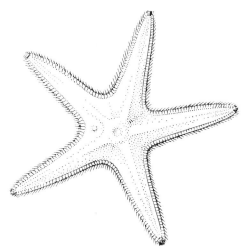

Sand burrowing starfish, *Astropecten irregularis*
(×¾)

which live in limy tubes fastened to rocks or shells and have little plugs for
closing the opening when the plumose feeding tentacles are withdrawn, are
common and so are the equally firmly attached saddle-oysters.

More active members of the epifauna include whelks, chitons and also an
unusual mollusc, the file shell (*Lima hians*). The body is fringed with long orange
tentacles which cannot be withdrawn but extend out into the water making this
the most beautiful of bivalves. It has the unusual habit of living in a nest, often
within the holdfasts of *Laminaria*, which it constructs out of its byssal threads.
This animal can also swim. Sea urchins are found here, also the common brittlestar (*Ophiocoma*), and many starfish such as *Asterias rubens*, with *A. glacialis*, a
related larger, grey species. These animals wander about in pursuit of food. In
cracks and holes live the sea gherkins, their whitish bodies hidden and only the
tentacles exposed. There is the usual abundance of crustaceans, such as the
lobster (Plate 41), and many crabs, including the edible species and several
spider crabs (e.g. *Inachus*, *Hyas*, *Macropodia*) which have delicate limbs and
disguise themselves with pieces of weed fastened to small hooks on the shell
and limbs (Plate 14). Prawns of various kinds are always to be found among
rocks.

The infauna comprises innumerable bivalves, most of them suspension feeders such as species of *Venus*, *Spisula* (p. 215) and *Mactra*; others, living in muddier bottoms with much organic matter, which are deposit feeders and include species of *Tellina*, *Abra* and *Gari*. Associated with the bivalves are carnivorous snails which bore into and feed upon them. Burrowing sea urchins, such as *Echinocardium* (Plate 14) and *Spatangus*, and the little heart-urchin, *Echinocyamus*, live in sand, and also a particular starfish named *Astropecten*, which has pointed instead of sucker-like tube-feet. It burrows in sand, capturing and swallowing small bivalves. Several kinds of worms habitually live in sand, while of crustaceans the commonest are the swimming crabs, which have the terminal joint of the fifth pair of legs flattened and paddle-shaped, enabling them to swim upwards. Especially adapted for a buried existence is the masked crab, *Corystes*,

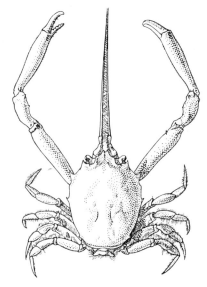

The masked crab, *Corystes cassivelaunus*, which lives vertically disposed, in sand. (Slightly reduced)

which lives with only the extreme tip of its long, upwardly directed feelers exposed. Each of these has two rows of stiff hairs which, when the feelers are placed together, interlock so that a channel is formed down which water can be drawn. Other devices for obtaining water for respiration while burrowing are the long siphons of the bivalves, and the rows of vibrating spines which keep up a circulation of water through the burrows of *Echinocardium* and other such urchins.

Burrowers are also typical of a muddy bottom. There are anemones of which only the mouth and tentacles appear above the mud, a variety of worms which burrow or else live in parchment-like tubes and also scavengers like the interesting sea mouse (*Aphrodite*), which attains a length of 15 to 18 cm and has a broad back covered with mouse-coloured felting, beneath which are scales, while from the sides of the body spring clusters of spines and beautiful iridescent hairs. The big sea cucumber or trepang (*Holothuria*) lives chiefly in mud, attaining to a length of 30 cm and a diameter of 5 cm, and has thick brown or yellowish skin. It moves, like its relatives the starfish, by means of five rows of tube-feet, and has the curious habit, when irritated, of shooting out masses of sticky white threads, which swell up greatly in water and may completely incapacitate attacking animals. Because of this habit, it is also called the cotton-spinner. On greater provocation it ejects its stomach and entire viscera, later growing a fresh set. Crustaceans are not especially common in mud but a number of small species occur.

Many of the animals just mentioned, and also a host of others, occur in the outer region of the Sublittoral which extends to the edge of the continental shelf. Indeed these regions contain an exceptionally dense and varied population of animals. The nature of the bottom continues to vary especially in the shallower regions where water movements still affect the bottom. With increasing depth the tendency is for a greater uniformity of mud. But there are wide areas for attachment of epifaunal animals. Reference can only be made to a few of the commonest or more interesting inhabitants. Suspension feeding and deposit feeding animals are commonest, with the latter increasing relative to the former in greater depths. With them are the inevitable carnivores. Plant life soon disappears, especially if the water is turbid. In very clear water, however, the limy coralline weeds, the red nullipores of which the commonest is *Lithothamnion*, may be abundant on rocky bottoms to depths of over 100 m.

Representatives of the Foraminifera, which live in limy shells often consisting of a series of spirally arranged chambers, are amongst the smallest members of the bottom fauna. On hard surfaces grow great masses of sponge, such as the yellow *Cliona* or else *Desmacidon* which produces endless amounts of slime so that when it is put in a bucket of water the whole becomes thick and sticky. Many crustaceans and worms burrow into these sponges or shelter in the crevices. Thick growths of hydroids and miniature forests of the pink coloured sea fans (Plate 13), the numerous branches of which extend all in the one plane, like a fan, are found on rock or stones. On muddy bottoms the most striking inhabitants are the sea pens (Plate 44), creatures allied to the anemones and hydroids, but consisting of a main, central stem, from the upper half of which spring branches bearing tiny polyps, the whole being distinctly reminiscent of a quill pen. Here also live the beautiful crinoids or feather-stars (Plate 5), another group of the starfish family, which rest on the bottom by means of a series of outgrowths from the underside of the body, while their ten delicate, foliaceous arms wave about in the water above. They begin life attached by a stalk, like the sea lilies of which we shall speak later, but afterwards break off and swim with graceful movements of their arms. The British representative, *Antedon bifida*, is red, brown or yellow in colour and one of the most beautiful members of the marine fauna. Many brittlestars and a variety of starfish, such as the fifteen-armed purple and rosy sun-stars (*Solaster*) (Plate 5), the red cushion-star (*Porania*) and the large yellow *Luidia*, occur in different localities, while burrowing urchins (Plate 14) and sea cucumbers are found on soft, and sea gherkins on hard, bottoms.

Amongst the many worms are included burrowers in mud such as the sipunculids, which have leathery bodies, one end of which can be drawn out into a long

proboscis, a feature also of the beautifully coloured nemertines, which have soft
and very extensile bodies, which often break into pieces when handled. The
bristle-worms are commonest of all, and include many of the kinds we have
already mentioned as well as others only found in this region. The most remarkable of these is the peculiar tube-dwelling *Chætopterus*, which lives in sand or mud
occupying a U-shaped parchment-like tube. In the middle of its body are three
broad segments which beat rhythmically—even when detached from the rest of
the body—and so maintain a steady current of water through the tube.

Commonest of all are crustaceans of all types. One of the barnacles, *Scalpellum*,
attaches itself to hydroids and is remarkable in that the large specimens are all
female, the males being minute creatures which live attached to the females.
Various ghost-shrimps and many members of the prawn and lobster family are
common on the bottom. Of the latter may be mentioned the Norway lobster
(*Nephrops*), the now well-known 'scampi', which burrows into muddy bottoms
(Plate 41), and the majestic rock lobster (*Palinurus*) with its handsome brown,
sculptured shell but without the large claws of the common lobster (Plate 41).
There are also squat-lobsters, such as the small *Galathea* (Plate 12) and the
larger *Munida*, which have long claws and broad, flattened bodies, the tail,
normally bent under the body, being straightened out and then suddenly flexed
when it is swimming. Various kinds of hermit crabs scavenge over the sea
bottom. Of the numerous crabs, there is the large, heavily armoured red spiny
spider crab (*Maia*) and, in more northern waters, the somewhat similar northern
stone crab (*Lithodes*), which is stone grey in colour and related to the hermit
crabs. There are many smaller kinds, some of which live in sand like the angular
crab (*Gonoplax*) which has a reddish-brown rectangular shell and eyes on the
ends of movable stalks.

A great host of molluscs find a home on the sea bottom, the most conspicuous
being large snails such as whelks, which include the common *Buccinum*, and the
smooth-shelled *Neptunea* with bold ridges running round its shell, the latter
coming from northern seas. The shell of these, and similar, animals is frequently covered with a moss-like growth which on examination proves to be a
very interesting hydroid called *Hydractinia*, which has three kinds of polyps,
respectively for feeding, reproduction and defence. A large sea slug, the triton
(*Tritonia*), is frequently found at these depths; it is pale coloured of various
shades and browses on dead-men's-fingers! The boat-shell (*Scaphander*)
ploughs its way through sand in the search for the small bivalves on which it
feeds. An occasional occupant of sand is the elephant's tusk shell (*Dentalium*),
sole representative of a distinct group of molluscs, which, as the name suggests,
have a curved tapering shell, open, however, at both ends. The empty shells
are frequently taken over by a small sipunculoid worm (in the north Pacific by
hermit crabs with straight bodies), which closes up the main opening except for a
small hole through which its long proboscis protrudes. The species of bivalves
are legion, from the large, very thick-shelled *Cyprina islandica* of the North Sea
to the many smaller kinds which are often almost incredibly abundant. Thus on
the Dogger Bank one such, named *Spisula*, occurs in patches often 80 km by 32
and at a density of anything from one thousand to eight thousand to the square

metre, while another, *Mactra*, is found in patches of 24 to 32 km in diameter and up to seven hundred per square metre. Everywhere abundant are the many different kinds of scallops, varying in size from the large *Pecten maximus*, up to 12·5 cm across, to species less than a tenth of this diameter.

Octopods occur here. There are two European species, the more northern lesser octopus (*Eledone cirrhosa*) (Plate 37) with a single row of suckers on the arms which occurs from the Mediterranean to beyond Great Britain, and the larger common octopus (*Octopus vulgaris*) with two rows of suckers. This is commonest in the Mediterranean, its most northern breeding area being off northern France from which, following favourable years, plagues of these animals may invade the southern shores of England. It also extends across the Atlantic into the Caribbean. Octopods live among rocks, darting out to seize their living prey. They usually crawl by means of the eight arms but are also graceful swimmers. body foremost, propelled by jets of water squirted out of the mantle (respiratory) cavity and directed by a tubular funnel equivalent to the crawling foot of a snail. The related but ten-armed cuttlefish (*Sepia*) (Plate 37) occurs on a sandy bottom and never crawls.

Amongst the animals we have not yet mentioned are various lesser known but interesting creatures. In some parts the sea bottom is covered with luxurious growths of the sea mats, the most conspicuous being the ross (*Lepralia*), which forms a massive skeleton consisting of many delicate limy sheets, while two others, *Cellaria* and *Flustra*, are frequently taken in great quantities in the dredge. Sea squirts of various kinds are found but are not so abundant as on the shore.

Bathyal Zone

The bathyal fauna is most easily studied in the Norwegian fjords which descend to depths of several hundred metres, the bottom being of mud inhabited by an interesting infauna, but the steep sides being of scoured rock to which a characteristic series of epifaunal animals is attached. These include masses of the stony coral, *Lophelia*, which forms branching colonies bearing delicate polyps like orange flowers on a yellow bush; with it are often great tree-like growths of the giant gorgonid (*Paragorgia*), which is bright scarlet. Sea pens (Plate 44) are common on soft bottoms, red, yellow or brown in colour, and varying in length between a few centimetres and a metre or more. Below 300 m may be found the big bivalve (*Lima excavata*), often 15 cm long and 10 or 12 cm wide, which has bright orange tentacles and flesh, the latter being considered a great delicacy by the fishermen. In similar localities the dredge brings up numbers of lamp shells or Brachiopods, representatives of an extremely old and formerly numerous group of animals which have a bivalved shell enclosing the body and are hardly to be distinguished by the casual observer from the bivalve molluscs. There are, however, many differences: the shell valves are unequal in size, are disposed on upper and lower sides of the body, not laterally, and have a different kind of hinge. The animal is attached to rocks by a small stalk which passes through a hole in the lower valve of the shell. Starfish and brittlestars are exceptionally abundant, the latter including the magnificent *Gorgonocephalus* (Plate 36), the long arms of which divide and sub-divide to form a writhing mass of tentacles like the hair of the Medusa it has been named after. Large stalked crinoids or sea lilies are conspicuous in these depths. Large red and brown sea cucumbers

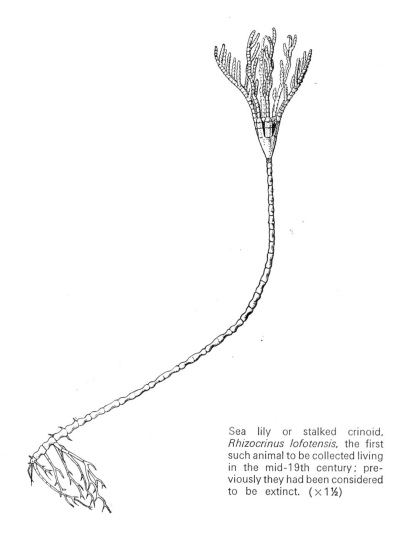

Sea lily or stalked crinoid, *Rhizocrinus lofotensis*, the first such animal to be collected living in the mid-19th century; previously they had been considered to be extinct. (×1½)

are very common, and also deep-sea sponges, some with root-like extensions for anchoring themselves in the mud, and others which grow on rock, one such, named *Geodia*, forming great rounded masses, often several metres across, dead white in colour and of strange shapes like the whitened bones of some prehistoric monster!

Abyssal Zone

Conditions in the Abyssal and the still deeper Hadal Zones are more uniform than in any other part of the world. There is absolute darkness, an unchanging temperature only slightly above zero centigrade and an enormous pressure amounting, at depths of some 6000 m, to around three tonnes per 6.5 cm^2. However, it has to be remembered that this primarily affects only animals containing gas, in lungs or, with fish, swim bladders. There is here no vegetation and animals have to be especially equipped for such a life. At one time it was thought that primitive animals might have survived in these depths and one of these, the little limpet-like monoplacophoran *Neopilina*, a survival from the Palaeozoic, was found in these depths by the Danish *Galathea* expedition. But most of

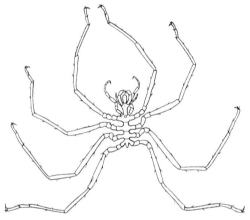

Sea spider, *Nymphon rubrum*, an intertidal example of a group of animals which are particularly abundant in great depths ($\times 2$)

these abyssal animals are highly modified, the reverse of primitive. Most of the bottom is of ooze and the place of attached animals is taken by creatures with long stalks which lift their bodies well clear of the bottom. Examples are provided by the graceful sea lilies, similar to the feather-stars of shallow water but with a long stalk in place of the short outgrowths on the underside, also by the sea pens, here at their most abundant, some alcyonarian 'soft' corals, with sponges, sea squirts and sea mats. Many of the crustaceans have long legs for lifting their bodies above the mud, and so have the sea spiders which may have limbs 60 cm long. There are relatively few animals with calcareous skeletons, molluscs, especially bivalves, being particularly rare, though some interesting examples of the latter, which are distinguished from their shallow-water relatives by becoming carnivorous, are found in the greatest depths. The total darkness has resulted, paradoxically, in some deep-sea animals losing their eyes completely, while others have developed especially large ones, sometimes on the ends of long stalks, like telescopes and probably capable of detecting any gleam of phosphorescent light.

Many deep-sea animals are of graceful and slender build, a result, probably, of the lack of rapid water movements, for there is no need for a strong skeleton and powerful muscles if currents and tides are absent. Striped and spotted animals, so numerous in shallower water, are remarkable by their absence, practically all deep-water animals being of a uniform colour, usually white, grey, black or red; blue and green animals are never found. Many of the fish have great jaws and powerful teeth by means of which they prey upon one another, other animals feed on the dead bodies which rain down from the surface kilometres above, while others again, such as the numerous many-coloured sea cucumbers, plough their way through the ooze, swallowing it as they go and extracting such nourishment as it contains.

Hadal Zone

This occupies the deep trenches in depths exceeding 6–7000 m. Here the sub-strate is of soft ooze with occasional masses of volcanic origin. The sparse, but characteristic population, consists largely of the recently identified pogono-phorans with echiurid worms, sea cucumbers (holothurians) and isopod crust-aceans; all feed primarily on organic matter in the bottom deposits. Represent-atives of other groups, notably decapod crustaceans, sea mats and brachiopods, have never been found in these profound depths.

Adaptations

Different animals are adapted for life under different conditions, and the bottom-living animals possess many structures and peculiarities which fit them for their peculiar life. We have already spoken of the manner in which burrowing animals maintain connection with the surface, while in Chapter 9 we shall see something of their devices for obtaining food. There is also the vital matter of reproduction. Very many of them produce eggs which hatch out into animals totally unlike their parents. Starfish and their relatives provide the most remark-able examples, but worms and crustaceans afford instances almost as striking (compare any adult crab with the larva crab shown in Plate 33). All such larvæ rise to the surface where they are for a time part of the drifting planktonic life described in Chapter 5. As already noted, on settlement to the bottom some may delay change to the adult form and assumption of the adult habit until a favour-able substrate is encountered. This mode of development ensures wide distribu-tion of the species concerned.

But in certain cases development is direct. The eggs are larger and the developing young assume the form of the parent usually before they leave the protection of its body. This is true of deep-sea animals where it would be im-possible for the larvæ to rise miles overhead and then sink back, and also of the inhabitants of shallow polar seas. There the period when larvæ can feed in the surface waters is so short that it is more efficient for the parent to store the needed food within the egg.

Bottom invertebrates do not make long migrations like the fish. This is impossible in the case of the many attached or rooted animals, while a large number of the others are too sluggish to move far. Octopods certainly move from place to place, sometimes causing great damage when they suddenly invade new areas, but this is probably the result of exceptional temperature conditions and is not a seasonal occurrence. Crustaceans, at any rate the larger ones, can move about freely; many, like the edible crab, coming inshore in the summer and then retiring into deeper water in the winter. In the summer the shore waters are warmer than the deeper water, but in the winter the reverse is true. Other

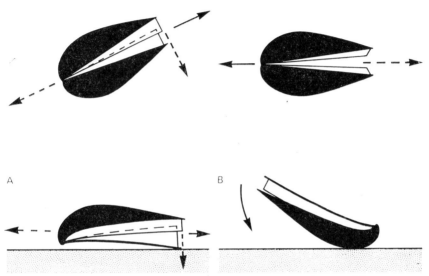

Diagrams illustrating swimming and other movements of scallops; solid arrows indicate direction of movement, broken arrows those of expelled jet streams of water. Top left, swimming movement, forward and upward; top right, escape movement, backward; bottom, righting movement of large scallop, *Pecten maximus*. A, movements when lying on flat valve, B, resultant action which brings it to the normal position with flat valve uppermost

animals which can move about are the scallops, especially the smaller 'queen' (*Chlamys opercularis*) which, by the continuous flapping of its shell valves, is able to swim about, hinge side hindmost (movement which is reversed when the animal is alarmed). The queen lives in large shoals which move freely about the sea bottom, changing their feeding grounds from time to time. Indeed there is a constant movement on the sea bottom, certain animals mysteriously disappearing from a given area and then as mysteriously reappearing, and we know nothing either of the cause or the extent of their movements.

Significance of the Bottom Fauna

Finally we might note the importance of this bottom fauna of invertebrate animals. Essentially these animals feed on the surface-dwelling plants, the microscopic members of the plant plankton. This comes to them either direct by turbulent water movements if they live in shallow water or more indirectly after death and some degree of decomposition if in deeper water or even under shallower conditions if water movements are slight, as over a mud bottom. Other members of this fauna are scavengers and there is the invariable assemblage of carnivores. But in general the invertebrate bottom fauna represents what is known as secondary production depending as it does on the primary production of surface-dwelling plants. In its turn it provides food for the bottom-living or demersal fishes such as the numerous members of the cod family, the flat-fishes and the skates, all highly important food fishes and representing tertiary production.

Chapter 4
SWIMMING ANIMALS

Over the sea bottom lie the waters of the ocean extending for kilometres in all directions. Above the surface of the land the air is the medium through which birds and insects, and even man, can make their way with great speed. In the same way the sea water forms a medium through which certain animals can travel rapidly from place to place by swimming. Although many minute animals can swim (in the true meaning of the word), that is, make their way through the water, it is usually agreed that the real swimming animals of the sea are those only who can make headway with sufficient speed to move from place to place against any current they may meet. Those smaller creatures that can swim without sufficient strength to stem the currents, and are drifted to and fro by them, are usually classed under the heading of drifting life, and will be dealt with at length in the next chapter.

Fishes

Most sea fishes can be divided into two main groups, the cartilaginous fishes or elasmobranchs, and the bony fishes or teleosts. In the group of cartilaginous fishes are to be found all the shark family—dogfish, tope, sharks, rays and skates. These are characterized by the fact that no true bone is to be found in their skeletons, but that they consist of that curious transparent substance known as cartilage, which, for all its delicate appearance, is in reality very tough. In the bony fishes on the other hand, the skeletons are all made of true bone, which is hard and brittle. This is only a further stage in evolution, bone being actually cartilage within which strengthening deposits of lime have been added. The skeletons of very young animals consist of cartilage, and it is only as they grow that the lime is deposited to build up bone. This being the case it is not surprising to find that the cartilaginous fishes are more primitive than the bony fishes, and appeared first in the course of evolution. In the struggle for existence during the history of the world, however, the fishes with true bony skeletons have been by far the most adaptable and are now very much more numerous both in kinds and in numbers than are the shark family.

But this is not the place to enter into a scientific discussion on the evolution of fishes or to describe in detail all the many different species of fishes that exist at the present day. It is rather the intention of these pages to show how some of the various fishes behave and live in nature and the part they play in the world under the sea.

Spawning Habits

In describing the life of any fish, it is reasonable to begin from the day of its birth. Most marine fishes, but not quite all, lay eggs; only a few are viviparous, that is, are born alive and do not hatch from an egg previously shed by the mother. Amongst these few are the viviparous blenny (*Zoarces*), one or two kinds of dogfish and the sawfish (*Pristis*).

But let us consider the majority, the egg-laying fishes. Some fish lay their eggs on the sea bottom attached to stones, shells and weed. The eggs of these fish when first shed are covered all over, or in certain places, with a sticky secretion, which soon hardens and glues the egg fast to the stone or shell with which it may be in contact. This method is typical of many of those little fishes

Left, eggs of blenny (× 25) ; right, eggs of goby (× 20)

which are so common in the rock pools and the tidal zone. The blennies, the gobies and the suckers all fasten their eggs in little clusters to the insides of empty shells, and under overhanging ledges or in crevices in the rocks.

The eggs, which are comparatively few in number, vary considerably in shape. Those, for instance, of the common blenny are rounded or spherical, with a little disc-shaped base that cements them to the rock, while those of the gobies are elongated and flattened.

It is quite a common occurrence for these kinds of fishes to guard their eggs. Not infrequently a blenny (*Blennius ocellaris*) may lay its eggs inside a bottle, and if this bottle is dredged up in a net it is quite likely that the parent will be found inside keeping watch over them. They have even been found inside an ox-bone.

The gunnel or butterfish (*Pholis gunnellus*) guards its eggs by coiling itself completely round them; the male stickleback and several species of wrasse build nests of weed within which the eggs are deposited. In the case of the pipe-fishes, which live amongst the sea grass, the males carry the eggs in a pouch under their tails.

Some shore fish, such as those mentioned above which lay their eggs in masses, or in nests, often have their own territories which they guard with aggressive behaviour against any intruder.

Almost all of our food fishes and many others, however, show no such parental solicitude for their offspring's welfare. After swimming up together in the spawning act they merely cast their eggs forth freely into the water, and these eggs are not heavy and sticky as those mentioned above, but are so nearly of the same weight as sea water that they drift about in the water layers at all depths between the surface and the bottom. Here they are at the mercy of tide and currents and are carried many kilometres from the region where their parents spawned. The dangers are many; they are not in sheltered crannies with parents to guard them; there are enemies all round to devour them; if their own parents happen to meet them again they will eat them with the greatest relish. What then is the provision for their safety? There is a well-known proverb, 'there is safety in numbers', and perhaps nowhere does this hold better than in the case of these carefree fish. We remarked above that those fishes who fixed their eggs to the rocks laid comparatively few; by this is meant never more than a thousand and probably considerably less. But what is this to the five million eggs a cod will lay, or the eight million laid by a turbot? And yet such are the dangers to be faced during the life of a fish that it is doubtful whether, at most, as many as ten out of these millions will survive to maturity.

These drifting eggs are to the naked eye like little glass-clear balls, varying in size from that of a small pin's head to about that of a radish seed. When first cast out into the water the eggs of many kinds of fishes are indistinguishable save by slight differences in size. But after a few days the young fish begins to form within the egg and there appear upon it flecks of colour, black, yellow or red, generally so disposed on the body of the fish as to make the different species quite distinguishable (Plate 8). There are, in addition, in many eggs, globules of oil which either by their size, number, or colour, make it at once apparent to which species they belong. Thus to the specialist the identification of some fishes' eggs becomes no harder than does that of the birds' eggs on land, the only difference being that it generally has to be carried out under the microscope.

The eggs of the angler fish (*Lophius piscatorius*) present a striking contrast to these single drifting eggs, being kept together in a large ribbon of jelly, which may float at the water surface. Egg masses of the angler fish have been reported a few square metres in extent, flat expanses of jelly carrying millions of eggs.

Practically the only one of our food fishes which does not lay drifting eggs is the herring. The herring's eggs are deposited on the sea floor, in certain localities only, on stony ground (Plate 39). The eggs are sticky and cling together in clumps in the crevices between the stones among which they fall. They are much enjoyed by other fish as food, especially by the haddock (*Melanogrammus æglefinus*); indeed, some of the spawning grounds of the herring have been located by the fishermen owing to the presence in their trawl catches of what they call 'spawny haddock', that is, haddock whose stomachs are packed full with the herring's eggs.

Many skates and dogfish have curious eggs, which are very familar to those accustomed to rambling along our beaches. They are the so-called 'mermaids'

E

purses', little horny capsules with spines or tendrils at the four corners. The smaller, narrower type, with long curling tendrils at its corners, is that of the dog-fish (*Scyliorhinus*), while the large, broad ones with the four spines belong to the skates. The dogfish eggs are attached to seaweeds by the parents, who wind the tendrils securely round the stem of the weeds; these eggs are occasionally to be found on weeds at low tide with the yolk and small developing fish in-side, because the 'mermaid's purse' is of course merely the case in which the egg lies. The skate's eggs, on the other hand, are deposited in deeper water, where they are buried in the sand with the two spines projecting above the

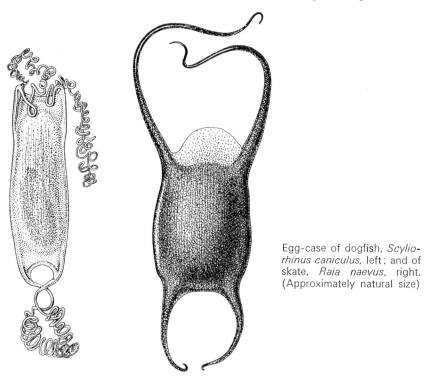

Egg-case of dogfish, *Scylio-rhinus caniculus*, left; and of skate, *Raia naevus*, right. (Approximately natural size)

surface, and through these a current of water is drawn into the case by the developing fish for breathing purposes. These eggs of the dogfish and skates take many months to hatch, although most other fishes hatch out within about a fortnight after the eggs are laid in cooler latitudes and within 24 hours in the tropics.

One of the most surprising cases of spawning instinct is exhibited by a little fish that abounds on the shores of California. This fish, the grunion (*Leuresthes tenuis*), belonging to the Atherine family, deposits its eggs in the sand at the high-tide mark along the sandy beaches. This it does one to three days after the highest spring tide, burying the eggs 5 to 8 cm beneath the surface of the sand. Here the eggs remain and develop, until in a fortnight's time, when the high tides once more cover them, they are ready to hatch. Now notice the providence of nature. If those eggs had not been laid just after the highest spring tide, there is quite a possibility that the next high spring tide, a fortnight later, might not have been quite so high, so that the eggs would not have been reached by the water; and in this case, being ready to hatch, they could not have survived another fortnight. Also, by being laid after, instead of before, the highest tide, they would be unlikely to be washed out of the sand before they had fully developed, since each successive high tide would be slightly lower until the spring tides started once more.

Early Life

Generally when the baby fish hatch they are extremely small. A young whiting (*Merlangius merlangus*), for instance, which is typical of many marine fishes, is only about 3·5 millimetres in length. It carries on its under-surface an oval sac of yolk on which it is nourished for the first few days of its free existence (Plate 8). When this supply is exhausted the young fish starts to feed. At this stage in its life the fish is drifting freely in the water layers above the bottom, and because of its small size and feebleness it can make no headway against the currents, but is carried along by them. All around is a community of other drifting animals and plants, and on these the little whiting makes its meals. After drifting thus for a fortnight or longer the fish begins to assume its adult character and appears as a true miniature of its parents. It is then about 2·5 cm in length and can swim with considerable agility, and seeks out new grounds in its search for food; some kinds of fish at this stage take to the sandy bottoms, while others seek the rocky coves and bays along the coasts. For the whiting, how-ever, a very interesting stage in its life-history has yet to be passed through. The little fish seeks out those big blue or red jellyfish, *Cyanea*, which abound at the same time of the year. These beautiful animals vary in colour from a deep brown red to the most heavenly ultra-marine blue. They can be found of all sizes, from that of a small mushroom upwards to 30 cm or more in diameter, and they have even been recorded in the cold northern waters to reach the immense size of 2·3 m across, with tentacles 37 m in length. These tentacles are armed with batteries of stinging cells. Under the shelter of these jellyfish the small whiting finds a temporary home. On a calm day it is at times possible to see one of these jellyfish floating near the surface of the water, and all around within a radius of a few metres can be seen numbers of these little whiting darting about, picking up their food. A sudden splash with the oar will drive them all beneath their curious shelter where they rest secure, trusting in the stinging powers of their host as a protection against their enemies. And the amazing thing is that the jellyfish allows it and does not attempt to capture them, although the red species, *Cyanea capillata*, can sting man quite severely and both species feed on plankton animals. In European waters the baby horse-mackerel (*Trachurus trachurus*) also seek this floating shelter, while in American waters young butterfish (*Poronotus triacanthus*) do likewise, as well as young haddock and cod from both sides of the Atlantic.

All the flat-fishes, such as plaice and soles, spend the first days of their lives drifting freely through the water. But at this period they are rather unlike their parents. The full-grown fishes, as we all know, have both eyes on the same side of the head; but not so the very young fishes, which are quite symmetrical, with one eye correctly placed on each side of the head. After two or three weeks, however, the eye on one side begins to move and slowly travels over the top of the head until it reaches the other side. At this stage the fish turns over on its side and seeks the bottom where it lies, with eyes both pointing upwards. This shows that many of the flat-fishes are flattened sideways, and are, indeed, living on the

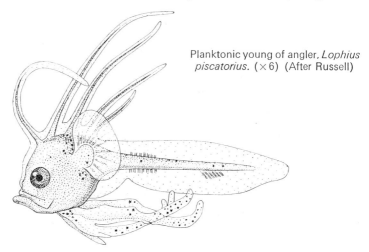

Planktonic young of angler, *Lophius piscatorius*. (×6) (After Russell)

bottom on their sides. Such fishes are the plaice, the sole, the brill, the turbot and many others. Some fishes, however, such as the skate and the angler fish, may be flattened from above downwards, and then they swim over the bottom truly on their stomachs.

The young stages of many fish are totally unlike the adults. The angler fish, for instance, so shapeless and cumbersome when grown up, is a most beautiful object soon after hatching. Its fins are drawn out into filaments of fairy-like delicacy. Like other young fish, the angler spends its early days drifting in the water layers above the bottom, and the long-drawn-out filaments of the fins help to keep it suspended in the water.

Migrations

It is obvious that, if, for several weeks in the early stages of their lives, most fishes are going to drift freely about at the mercy of tide and current, they will be carried far from where their parents shed the eggs. The greatest importance therefore attaches to the position of the spawning region, from which after a definite time the young fishes must have been carried to suitable grounds for feeding and growing. For this reason many adult fishes undergo spawning migrations, that is, they move off all together to a chosen locality to shed their eggs. The plaice, for instance, in the southern North Sea move farther south so that their eggs and young are drifted by the prevalent currents on to the so-called 'nurseries' in the shallow, sandy-bottomed regions along the Dutch and Danish coasts (see p. 212). The herring, too, move in from deeper water to deposit their eggs on the bottom in certain regions, and it is then that the fishermen set to work to catch them in drift nets or purse seines.

But most surprising of all spawning migrations is that of the common freshwater eel. For years men in all countries of Europe wondered how it is that the eels in our streams thrive and multiply, and yet nobody had ever seen their eggs or tiny young. Izaak Walton in his *Compleat Angler* says: 'Some say that they breed, as worms do of mud; as rats and mice, and many other living creatures are bred in Egypt, by the sun's heat when it shines upon the overflowing of the river Nilus; or out of the putrefaction of the earth, and divers other ways . . .' 'and others say, that as pearls are made of glutinous dewdrops, which are condensed by the sun's heat in those countries, so eels are bred of a particular dew, falling in the months of May or June on the banks of some particular ponds or rivers . . . which in a few days are, by the sun's heat, turned into eels.'

But just as we know that pearls are not made from dewdrops, so are we now certain that eels do not spring from mud, or putrefaction, or drops of dew. For in the year 1922 a great Danish oceanographer, Dr. J. Schmidt, found the true spawning place of the eel and cleared up the mystery once and for all. He disclosed the amazing fact that when the eels in our rivers become mature they set off on one of the longest journeys as yet known to be undertaken by any fish.

When the common freshwater eel of our rivers has reached a length of about 30 cm, it changes its appearance and puts on what is known as its 'spawning livery'. Instead of its usual yellow colour it becomes quite silvery in appearance and hence at this stage is known as the 'silver eel'. Its eyes enlarge and become adapted for the dim light of the deep waters of the ocean. In this condition the eels work their way down to the mouths of the rivers. In the late summer and autumn this migration takes place and the eels push on out of the estuaries and into the open sea. Then starts the long journey out into the deep water and so into the Atlantic Ocean. Of the speed at which the silver eels go on their journey we know little. Once into the deep water they become lost to our observation. In the Baltic they have been captured and marked. Their recapture on their way to the North Sea shows that they had travelled at rates of about 12 to 52 km a day, but they were only a very short way on their total journey, for the next we know of them they must have arrived in the deep central part of the North Atlantic, known as the Sargasso Sea (a distance of between 3000 to 5000 km), although none of the spawning eels have themselves been seen there. There is, however, irrefutable evidence that they must have been to those regions, because at the end of winter and during spring the baby eels are there. The young eels are very unlike their parents; in fact so unlike that the first time one was seen it was considered to be a new species of fish and given the name of 'Leptocephalus'. They vary from 6 to 76 mm in length according to their age. They are flattened sideways so as to resemble a leaf, and are quite transparent (Plate 15). These baby eels now start on the return journey and we can in this case get some idea of the rate they travel, for Dr. Schmidt, by measuring very large numbers, showed that they take three years to reach the European shores. During the long journey across the Atlantic they grow, and it is by their growth that their birthplace has been located; catches made between our shores and the central Atlantic exhibit these eel larvæ in ever decreasing size. Near the coasts they are about 76 mm long, but down in the locality of their birth they are only 6 mm in length, and indeed, in this region, eggs have been taken that might perhaps be those of

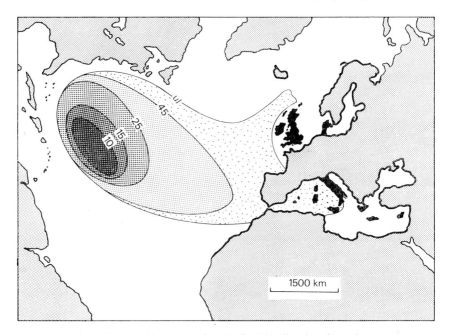

Distribution of larvae (dotted area) and of adults (black stripes along coasts where species occurs) of the European eel. The contours show the limits of occurrence of the different sizes, i.e. larvae less than ten millimetres long have only been found inside the ten-millimetre curve. ul is the limit of occurrence of unmetamorphosed larvae. (After Schmidt)

the parent eels. The eggs are about the size of a pea, quite transparent, and drift in the water layers at depths of about 200 m, just like many of the other fishes' eggs mentioned earlier in this chapter.

When the eel larvæ have reached their destination on the coasts of Europe, a very remarkable change comes over them. From their leaf-like shape they gradually assume a true eel-like appearance, becoming narrower, rounder, and at the same time slightly shorter in length. After this change has taken place they are the typical little eels, known as 'elvers', that ascend some rivers in such countless numbers. Here they remain in fresh water, feeding and growing. After a course of time, that may vary from five to twenty years, the eel puts on its silver spawning dress and once more sets out on the return journey to that deep part of the Atlantic where it spawns and probably dies.

There is in America an eel which is very similar in appearance to the common European freshwater eel, differing only in certain minute anatomical details. This eel undergoes a similar life-history to that just described, with the only difference that the spawning ground is slightly to the west of that of the European eel, and the leaf-shaped larvæ take only one or two instead of three years before they are ready to ascend the American streams as typical elvers (Plate 15).

This difference in the life-histories of the European and American eels is of great interest in showing the speeding up or slowing down of development to suit the environment. If the larvæ of the European eel took only one year to reach metamorphosis they would still be far from their destination on the European coasts, when they had assumed the shape at which they normally ascend the rivers; while the American eels by a reduction of the period of larval existence are ready for metamorphosis by the time they have arrived at the American coasts.

A somewhat similar, but reversed, life-history is that of the Atlantic salmon (*Salmo salar*) whose adults undergo most of their growth in the sea, and then move up into the rivers to lay their eggs. The young live for one to three years in the fresh water before migrating down to the open sea to feed on the large food supplies available there and to make that very great growth that distinguishes them from their relatives the trout, who spend all their lives in the rivers.

The feeding grounds of the Atlantic salmon are in northern waters and since 1960 a fishery has been developed for them off the coast of Greenland.

The different species of salmon which spawn in rivers of lands bordering the North Pacific have their feeding grounds in the open waters of the northern part of that ocean and these commercial salmon fisheries extend over a large area (see also p. 9).

Distribution

It can be seen that because of the large journeys carried out by many fishes, the area of their distribution must vary at different times of the year and be rather widespread. Nevertheless there is noticeable, in the seas of the world, a definite zoning of the distribution of different species of fishes. There are, of course, differences in distribution to be found between such fishes as live always in the tidal zones or in the vicinity of rocky coasts, as opposed to those of more open water. Such differences can be noticed in comparatively small areas; but there are, as well, differences to be found over very large regions. Everyone knows that if he visits tropical latitudes he will see fishes that he never sees in more northern waters. A visitor to such widely separated fish-markets as those of Grimsby in England, Trieste in Italy, and Colombo in Ceylon, will at once be convinced of this. Each market will display its own characteristic fish, and it is doubtful whether any fish would be found common to all three markets. Such localization of faunas is easily understandable on land, where such barriers as deserts, high mountain ranges or large areas of water limit the different animals to their own special regions. But in the sea, with all the oceans of the world connected, there are no such purely mechanical barriers present to limit the dispersal of the different kinds of fish and prevent them from presenting a perfectly uniform distribution from ocean to ocean and sea to sea. How is it then that one finds nevertheless that certain widely separated areas each possess their own characteristic population of fishes ? It is evident that there must be some barrier, and a general exploration of the chemical and physical properties of sea water has shown that in all probability the chief factor in the distribution of the fishes is the temperature of the water.

It is generally to be noticed that in northern waters, wherever the temperature of the water is less than ten degrees centigrade, typical northern fishes will be found, such as cod, halibut, haddock, herring, and many others. South-western coasts of the British Isles, such as those of Devon and Cornwall, are on the southern limit of the distribution of these northern fishes; and, at the same time, here occur the northern limits of certain southern warm-water-loving species. It is, so to speak, a meeting ground of the two great areas and certain fishes common to both occur. Besides the herring and the cod, both northern representatives, are, for instance, the pilchard, and occasionally the red mullet and the anchovy, fishes which are typical of the warm waters of the Mediterranean and Atlantic. A similar demarcation area is to be found off the coast of California where the sardine meets the more southern anchovy.

Within the large areas, also, it is to be noticed that there are further subdivisions; certain fishes which live in the very cold part of the Norwegian Sea, and in the Arctic waters, are characteristic of those regions.

In the tropics, again, the fish population is characteristic, and here, especially in coral reef areas, are to be found many of the most brilliantly coloured fishes in the world.

While the distribution of different kinds of fish in the water is thus largely governed by temperature, large changes in geographical distribution may be brought about by alterations in climate. For instance, after the 1920's there has been an amelioration of the climate in the far north. This resulted in a northerly extension of the warmer waters bringing with them such fish as herring and cod into areas in which they were previously absent. This gave rise to new fisheries in the north, and of course this extension of habitat occurred with other animals as well as fish (see p. 224).

Such changes also resulted in warmer-water species moving farther north, the pilchard for instance replacing the herring in some European waters and the anchovy taking the place of the sardine in the eastern Pacific.

There are indications that the glaciers which receded during this warm period are once more advancing and that the situation just described is now being reversed.

In the geographical distribution of fishes, apart from these differences caused by temperature, there are also differences that are occasioned by the depth of the water. There are shallow-water fishes, deep-water fishes, and abyssal fishes, that is, fishes who live on the very flat plains in the deepest parts of the oceans forming the abyssal zone described on p. 49.

It is natural that those fishes which will be limited to certain areas by depth boundaries are those that live most of their time swimming on, or close to, the sea bottom itself. Fishes such as herring, mackerel, sardine and tuna, so-called pelagic fishes, can have a much wider area over which to roam, because, swimming as they always do in the upper water layers, they can keep to the depths they most prefer, irrespective of at what depth the actual bottom of the sea may be.

As examples of fish whose distribution is limited by depth, we can name the plaice, which lives in the comparatively shallow flat areas such as the North Sea; the hake, that roams along the deep-water shelf from 200 to 900 m, that constitutes the continental slope; and that curious fish of the cod family known as

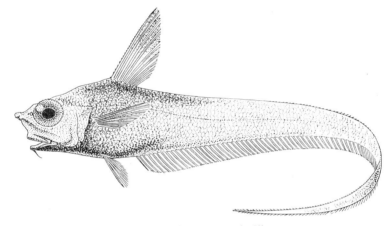

The rat-tail, *Nezumia*. (×⅔)

the *Nezumia* or rat-tail, which spends the greater part of its life in the cold dark depths of the ocean.

These remarks on the distribution of fishes will help to explain also the distribution of some of the major sea fisheries. Certain fish, such as the mackerel (*Scomber scombrus*), which prefer warm water, will be found around the north coast of Scotland and also in the North Sea. This, at first sight, appears rather contradictory, but the explanation is simple when we realize that there lies the course of the North Atlantic Drift, and it is to the warm waters of this oceanic current that the mackerel are keeping. Yellow-fin tuna tend to congregate near oceanic fronts at 24°C where cold and warmer waters meet and cause concentration of food organisms.

It is evident that all the fishes that live most of the time on the bottom will be limited by the depth barriers to coastal waters or the regions of comparatively shallow banks. There are, however, many fishes practically unknown to most people that may be termed oceanic. These fishes roam about all through the water layers out in the open oceans. They are mostly very small, the largest being a few centimetres in length. Many of them are very bizarre (Plate 20) in appearance and possess most interesting organs for the emission of phosphor-escent light (see Chapter 8).

The distribution of these small oceanic fishes is also of great interest. They, like most other fishes, are apparently limited in their geographical distribution by those unseen temperature barriers. At the same time just as the bottom fishes appear to be restricted to areas within certain depths, so in the open waters of the ocean these fishes are to be found living, each species inside a definite limited range of depth below the sea surface. There will be, for instance, those that occur mostly between the surface and 200 m; others, again, will never be met with in these water layers, but are only found below perhaps 400 m; and yet others may be captured only at very great depths, such as at 2000 m. The depths at which many of these deep-sea fish live is determined by light intensity for temperature differences are slight.

Shape

To the average person the word fish summons up in the imagination a silvery, wriggling, slippery animal of a certain definite torpedo shape.

This characteristic spindle or torpedo shape is admirably suited to the habits of most fishes. If we take, for example, a fish like the mackerel, we realize that not only is it beautiful and elegant, but that it is efficient; the fish is made to swim and nature has made no mistake. Just as man has learnt many lessons from the forms of birds to aid him in the designing of the most efficient aeroplanes, so can much be learned from the mackerel about the best shapes for rapid motion through the water. Its shape is 'streamlined'; that is, it is adapted so that in its progression through the water the least possible friction is set up and it can cleave its way without hindrance. If a square block of wood is placed in a flowing stream there will be considerable resistance on the front surface and at the same time many eddies will be created just behind it, which act as a suction and im-pede its passage through the water. If now the front surface be rounded off the water becomes parted with a minimum of resistance; the water can now be seen to flow round the block in two streams which meet again some little way behind. Between the back surface of the block, the two surfaces of the divided stream, and the point at which they meet, is a spindle-shaped mass of water full of eddies. If the block be built up behind with wood of such a shape as just to fill this eddy-ing area, then it will be found that the shape produced sets up the least possible resistance in its passage through the water. Its form is exactly the same as the outline traced by the water flowing round an object and hence the term 'stream-line'. It is just this shape that is typical of most fishes. In eel-like fish forward motion is imparted by muscles producing successive waves passing down the whole length of the fish. In fish like the salmon this wave motion is restricted to the tail region.

Slight variations in outline will be required to obtain greatest efficiency at different speeds, and this probably accounts in part for the variety of form shown by different kinds of fish. A cod, for instance, by the nature of its habits does not require to swim so continuously as a mackerel. A close examination of the mackerel shows that besides having this definite shape designed for speed, there is a smoothing off of all excrescences that may set up resistance. The bones around the mouth and jaws, that in many slow-moving fish are somewhat projecting, are here inset absolutely flush with the smooth outline of the fish. The crescent-shaped tail also, swept back like the wing of an aeroplane, is the most efficient shape for fast speeds.

Actual measurements of swimming speeds have now been made for some kinds of fish. Fish cannot swim as fast as is popularly supposed. For instance the

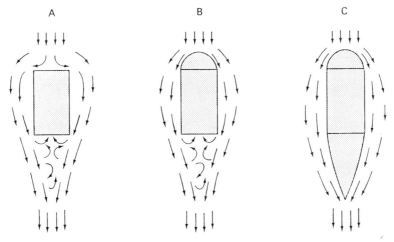

A, B, and C are diagrams showing the successive building up of an oblong block into a streamline shape to reduce resistance caused by eddies

speed that a herring can sustain for a long period is under 6 km per hour, and the maximum speed for a mackerel is only about 11 km per hour. Fast swimming fish like the tunny may reach speeds of up to 72 km per hour.

Very different in appearance are the fishes that live a great part of their time actually on the sea bottom. Many of them are quite flattened, so that they have a very large surface on which to lie, and their fins are so arranged that by a slight flapping they can throw up sand and pebbles which settle down on the broad surface of the fish's back and eventually completely bury it from view. The common sole and the plaice are typical examples of flat-fishes. They are remarkable in that in reality they are lying on their sides. As has been mentioned above, in their very youngest stages they are quite normally symmetrical like any other fishes, but at a certain stage, when they are between 13 and 19 mm in length, a curious twisting takes place in their head region and the eye from one side travels round so that eventually both eyes come to lie on the same side of the head. The fish then turns over and settles on the bottom on its eyeless side. In some species it is the left eye which changes its position and in others the right eye, and it is rather unusual to find a fish with both its eyes on the opposite side to that typical for the species.

Other fishes, however, are flattened quite normally. The skates for instance are flat from above downwards and the eyes keep their same relative position throughout life.

The gurnards are literally able to walk over the bottom. Just in front of the big fan-shaped pectoral fin on each side are three curved rigid spines, and if the fish be watched in an aquarium it will be seen that when moving over the bottom it is actually walking on the spines. The spines really belong to the fin; they are fin-rays which in the course of time have broken away from the actual fin itself and evolved into these curious legs, which have a tactile and tasting purpose.

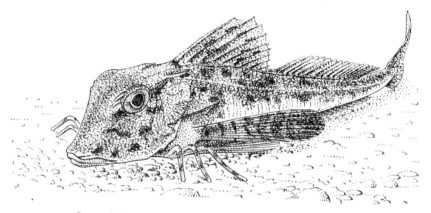

Gurnard, showing the separated fin-rays of the pectoral fin

Another way in which fins have been modified, is to form suckers by means of which a fish can hold on to a solid object. The common sucker (*Lepadogaster*) found in rock pools has a suctorial disc formed by a pair of fins on its under surface. In this case the resemblance of the sucking organ to the original fins can still be seen. But any fin-like appearance is completely lost in the case of the sucking disc of the *Remora*. This fish bears upon its head an object that looks more like the rubber sole of a sand-shoe than anything else. It is a very powerful sucker and with it the *Remora* fastens itself to the bodies of sharks and so gets rapidly transported from place to place and incidentally is able to pick up any remnants of food that the greedy host may drop. That this sucking disc is really a transformed back-fin has been discovered by an examination of the very early stages of the fish; when it is only about 13 mm long it possesses a normal dorsal fin like any other fish, but as it grows this fin gradually moves forward until it comes to lie on top of the fish's head, and in the meantime its structure is altering to that curious ridged form which acts as such an efficient sucker. The sucking action of this disc has been disputed, and it is thought by some that the clinging effect is brought about chiefly by friction.

The habit of the sucking-fish of fastening on to other fish has been taken advantage of by man, who actually makes use of the *Remora* to catch other fish. The method is practised by natives of such widely separated regions as the

Caribbean Sea, Chinese waters and the Torres Strait. Although being used for catching certain fish such as sharks they are chiefly used for capturing turtles. The sucking-fish has a thin line attached to its tail. On sighting a turtle the natives row up to within fairly easy reach of it and then throw the *Remora* towards it. The fish immediately swims to the turtle and attaches itself to its under-surface. By pulling on the rope, the boat and turtle can then be brought close together and the turtle captured. At times, however, the turtle dives, and the *Remora* does not stick fast enough to allow sufficient strain to be put on to the rope to lift the turtle which clings hard to the bottom. In such circumstances the native locates the exact position of the turtle by means of the string and then dives down himself and secures it with a rope. The practice was studied in the region of the Torres Strait very thoroughly by Prof. A. C. Haddon, who wrote: 'The sucker-fish is not used to haul in the large green turtles; I was repeatedly assured that it would pull off, as the turtle was too heavy; but small ones were caught in this manner.' He goes into full details of the attachment of the leash to the fish and adds, 'I was informed that in leashing a sucker-fish, a hole is made at the base of the tail-fin by means of a turtle-bone and one end of a very long piece of string inserted through the hole and made fast to the tail, the other end being permanently retained. A short piece of string is passed

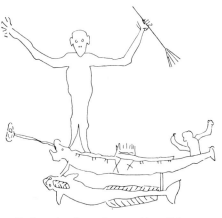

Native drawing of a sucking-fish, or *Remora*, attached to a canoe. (After Haddon)

through the mouth and out at the gills, thus securing the head end. By means of these two strings the fish is retained, while slung over the sides of the canoe, in the water. The short piece is pulled out of the mouth of the fish when the turtle is sighted and the *gapu* is free to attach itself to the turtle.' And after this, according to Prof. Haddon, the sucker-fish is eaten at the end of the day!

A very curious modification of a fin-ray is to be found in the common angler fish or fishing frog (*Lophius piscatorius*). In this case the front ray of the back fin is very elongated and bears at its tip a little tuft of filaments. The fish is truly named an angler fish for by means of this fin-ray it angles for its prey, which is lured on by the worm-like waving filaments. Even more curious are the near relatives of this fish which inhabit the dimly-lit layers of the ocean from 400 to 2000 m deep, the deep-sea anglers. In many of these the little lure at the end of the rod is phosphorescent and lures the unwary to their death (Plate 20), as did the wreckers on British coasts in days gone by. We can well imagine that, owing to the tremendous area over which these deep-sea anglers can roam in the open ocean, the chances of a male and female meeting are rather small. To overcome this there is a most remarkable provision. Females have been found carrying

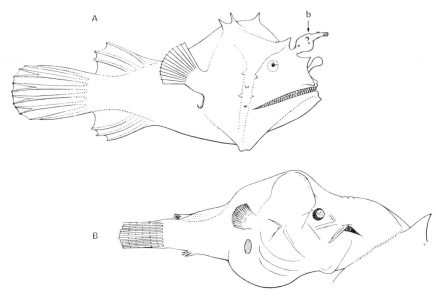

A. Female deep-sea angler, *Photocorynus spiniceps,* with dwarf male attached to head (b). (×1⅔) B. Enlarged drawing of male showing attachment (×6)

fused to their bodies tiny dwarfed husbands. It seems probable that while the males are still fairly numerous soon after they are born, before they have had time to be thinned out by their enemies, they fasten on to the first lady they can find, and there they stay for the rest of their lives. They become completely fused to their partners and lose all their individuality, becoming merely appendages on their wives!

In the flying fish (*Exocœtus*), it is the pectoral fins that have become modified to offer an enormously large surface to the air, and so act as gliders when the fish

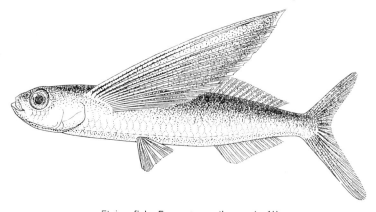

Flying fish, *Exocœtus spilopus.* (× ⅓)

leaps out of the water. Observations made by high-speed photography confirm earlier statements by E. H. Hankin that, if the initial leap does not carry the fish clear of the water the enlarged lower lobe of the tail is wagged rapidly so that by sculling action the speed is increased until suddenly the fish leaves the water. The fish remains in gliding flight with its wings held out stiffly, and the wings are not normally fluttered. Flying fish glide only a few centimetres above the water, unless lifted by an updraught of air. Observations have shown that 75% of the flights are into the wind. After a certain distance momentum is lost and the fish may sink towards the water until the lower lobe of the tail can

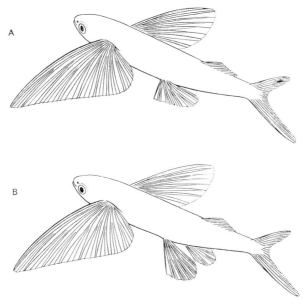

Showing position of fins of flying fish. A. When in full flight and B. before re-entering the water, the pelvic fins being depressed to reduce the speed of flight.
(After Hankin)

once more give the necessary propulsion for continuation of the flight. In a 26-second flight there were for instance 10 touch-downs. The longest recorded flight is over 30 seconds, during which time the fish may travel 305 m in the air. The average flight lasts 10 seconds or less, and the air speed is 48 to 64 km per hour. Such flights cover 30 to 90 m by tail wagging after leaving the water at about 27 km per hour.

In the sea horse the fins are often reduced to mere vestiges or are absent altogether. The fish is covered with a hard, jointed armour, and has a prehensile tail. There is a near relative of this fish which lives in Australian waters (*Phyllopteryx eques*) that is a piece of living camouflage. All the knobs and spines on its body are prolonged into leaf-like filaments, which give the fish exactly the appearance of the brown seaweeds amongst which it lives.

The torpedo, or electric ray, a member of the skate family, is remarkable for possessing on either side of its head two organs capable of generating quite powerful electric discharges. The electric organs are composed of muscular tissue in the form of innumerable small cells vertically arranged, the top surface of the organ being electro-positive, and the bottom electro-negative. A large fish of this species can give a shock sufficient to paralyse temporarily the arms of a strong man. At least one species occurs at times on the British coasts, but the majority live in warmer seas (see also p. 155).

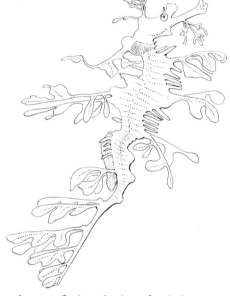

A sea horse, *Phyllopteryx eques*, with filamentous outgrowths for camouflage. (Natural size)

Some fish are noteworthy for their poisonous flesh and others for their powers of stinging. Poisoning from eating fish has long been known among the islanders of the tropical Pacific, but it was brought much to notice during the last war when many of the armed forces were present in the area. As a result much research has been done since on the subject.

A surprisingly large number, as many as five hundred species, of fish are known to be poisonous when eaten, and these are mostly coral reef and shore fish living around the islands. Their poisonous nature may be due to substances normally present in certain tissues, or more often to substances arising from organisms eaten by the fish. Although deaths occur each year from this cause, fortunately the dangerous species are known to the local inhabitants. Typical of such fish are the puffers or toadfish, *Tetraodon*, which have the curious power of inflating their bodies with air or water.

More unpleasant are those fish which have the power of stinging. Such fish are furnished with poisonous glandular tissue and spines with grooves down which the venom can flow. The spines may be present on the operculum, or gill-cover, and in the fins, or in the tail. Common examples of such venomous fish are weaver-fish, stone-fish, and sting-rays. The lesser weaver-fish (*Trachinus vipera*) occurs on sandy shores and has spines on the operculum and in the dorsal fin. The fish bury themselves in the sand and can inflict painful wounds when trodden on. Bathing for some time in hot water will relieve the pain. The sting-ray has a large toothed spine situated near the base of its whiplash-like tail, and used for defensive purposes. These fish also often lie buried in the sand, and if trodden on will thrust their tails upwards, thus driving the spine into the foot or leg. The most venomous fish known is the stone-fish (*Synanceja horrida*) a very unpleasant looking and well-camouflaged fish which frequents coral reefs, lying buried in the mud or sand, or on the rock or coral. This has spines in all its fins, eighteen in all, and bad stings may cause death in several hours.

But dangerous as the stone-fish is, by far the most venomous of all marine animals is a jellyfish belonging to the group popularly known as 'sea wasps' whose name is *Chironex fleckeri*. This jellyfish is transparent and about 23 cm in height. It frequents the bays in the tropical regions of Queensland in Australia. A bad sting by one of these jellyfish can cause death within three minutes.

Much could be written about different aspects of the behaviour of fish, but space does not allow. Like birds, nest-building fish and indeed many others have ritual behaviour during courtship. Special mention should also be made of another curious habit on which there has been much recent study. This is the habit among some species, such as small wrasses, of cleaning other fish. The small fish will remove and eat parasites from the larger fish, even going under the gill covers and into the mouth to do so. The cleaner will have a recognized station and other larger fish will come to be cleaned and in the process there is a well-defined ritual behaviour. Such behaviour is shown by a number of coral reef fish.

Whales

Whales are not fishes. They belong to the mammals, that large group of the animal kingdom the great majority of which live on dry land. Whales are in fact four-footed animals that, in the long course of evolution, have taken to the water for their permanent abode. Their external structure has, however, become so changed that they are almost fish-like in appearance, having the typical spindle shape so characteristic of fast-swimming fish. They possess two large flippers where their front legs or arms should be. Many of them carry a vertical fin on their back and their tails are developed into powerful flukes; but unlike those of fish, these tails are flattened horizontally instead of vertically, probably to aid in the continual journeyings to the surface to breathe. An examination of the skeleton reveals the fact that the front flippers are undoubtedly the same as a true leg, for buried in the flesh are typical leg bones ending in numbers of small bones similar to those of our hands or feet. But all these bones are completely buried and there is no external division of the flipper into arm, hand, and fingers. Although, externally, there is no sign of any hind limbs, yet in some species of whales, buried in the body in just the region where the hind legs should be, are to be found small remnants of bones much reduced in size and serving no

F

purpose—the last vestiges of the back legs of the land mammal, from which through countless ages the whales have been evolved.

Hair is one of the chief characteristics of mammals. It is essential for helping them to keep the warm blood that flows through their veins at the correct temperature. Yet a whale is practically hairless save for a few fine bristles in the neighbourhood of the mouth, which may or may not persist throughout life. Living as it does in the cool waters of the ocean it must therefore have some other means of preventing its warm blood from cooling, and to this end its whole body is encased in thick coatings of fat—the blubber.

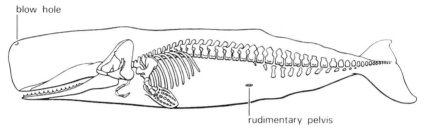

Outline and skeleton of sperm whale or cachalot, *Physeter catodon*

The whale's nostrils are situated on the highest part of the head, and it is through these that it spouts or blows. The well-known spouting of the whale is nothing more than the act of breathing. The air in the lungs becomes heated and steamy and is ejected with considerable force when the whale rises to the surface, and condensing in the cold atmosphere it appears like a spray of water.

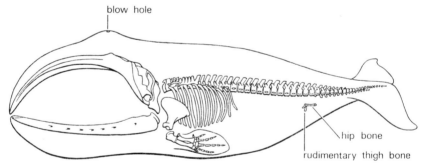

Outline and skeleton of Greenland right whale, *Balaena mysticetus*

Occasionally the whale starts to blow before it has actually broken the surface, in which case water is also driven into the air. Any water which has accidentally reached the lungs is also expelled with the breath.

There are many different kinds of whales, but they may be divided into two main groups, the toothed whales and the whalebone whales.

The great sperm whale is a good example of the toothed whales (Plate 18). Although there are no teeth in the upper jaw of the sperm whale, or cachalot, the lower jaw is well armed with twenty to twenty-five pairs of very sharp teeth. These teeth are most necessary for it feeds on a very formidable prey, the giant cuttlefishes or squids. Judging by the taste of the small common squid which is eaten on the Continent, the whale must feed on choice morsels, but it has to capture a fearsome monster before it can enjoy its meal. For some of these squids are some metres in length, with tentacles 9 m long or more, armed with many powerful suckers. It needs a monster like the whale to tackle them, and many whales are caught with great scars upon their bodies that bear witness to the mighty struggles that have taken place in the deep waters of the ocean.

Among other toothed whales the killer whales (*Orcinus orca*) are of interest, because they hunt in packs, feeding on fish and seals and even harrying the larger whales themselves. The common porpoise and the dolphins (Plate 19) are also representatives of this group, the porpoises being among the smallest of the toothed whales.

The whalebone whales do not possess typical teeth. They have instead long, thin plates of a horny texture, set close together all around the upper jaw. This is the baleen, or whalebone, that was once so fashionable in ladies' attire, but is now largely used to produce bonemeal. Each plate of whalebone is frayed out along its inner edge to form a mass of felted fibres (Plate 19).

With such teeth the whale obviously cannot prey on any powerful animal. It cannot bite or chew. Instead it feeds on the hordes of minute animals from 4 mm to 25 mm in length, that drift in countless myriads in the waters of the sea and go to form part of that drifting community known as the plankton. Opening wide its mouth the whale takes in a great gulp of water with all its contained living animals. Then closing its mouth it raises its tongue and forces the water out through the whalebone sieve and the animals which remain behind are swallowed.

The right whales are a typical example of this group, as are also the humpbacks and rorquals. Of these the blue whale, or Sibbald's rorqual (*Balænoptera musculus*), may reach the immense size of 26 m or more in length (Plate 18). The weight would be over 305 t.

Whales are distributed all over the oceans of the world, but while the toothed whales are to be found roaming throughout the central parts of the oceans where their prey, the squids, occur, the whalebone whales are chiefly confined to the cold Arctic and Antarctic waters and the coastal banks, where the drifting life on which they feed is most abundant. Many whales are stranded, dead or alive, from time to time along the coasts of the British Isles.

One of the more striking discoveries of recent years has been that dolphins, and probably other species of whales, are like bats capable of echolocation. By emitting high-pitched sounds they are able to avoid obstacles placed close together in the dark, and they can in this way also locate and identify their food. These investigations were made on the bottle-nosed dolphin (*Tursiops truncatus*) which is kept in captivity in large oceanaria. Dolphins are most intelligent animals and can be trained to perform tricks with evident enjoyment. There seems little doubt that whales can communicate with one another by sounds over long distances.

Seals

Other mammals that have taken up their abode in the sea are the seals, the sea-lions and the walruses. Unlike whales, which are unable to leave their watery home, the seals can move with considerable rapidity on dry land, although their gait can hardly be said to be gainly. Indeed all seals resort to dry land to bring forth their young, whether it be on sandy beaches, on rocks or on ice floes, and the young seals are said to be reluctant at first to enter the water.

There are many different species of seals, but only two are usually to be seen around British coasts, the common seal (*Phoca vitulina*) and the grey seal (*Halichœrus grypus*). Neither of these furnishes the beautiful sealskin of commerce, this product being obtained only from certain species known as 'fur seals' or 'sea bears'.

Seals feed mostly on fish, and are commonly a nuisance in the estuaries of some of the Scottish salmon rivers, where they do much damage to the net fisheries for the salmon. Some species feed on shellfish and other small marine animals. The walrus uses its long ivory tusks for digging in the sand and gravel to turn up the shellfish on which it feeds.

Common seals and grey seals usually stay submerged for only a few minutes, but can if necessary stay under for over twenty minutes. Although seals have been reported as diving as deep as 140 m experiments have shown that they cannot survive after five minutes at such depths. When diving, seals close their nostrils after breathing out to reduce the amount of air in the lungs as a precaution against the 'bends', and the blood supply to some organs is cut off. Weddell Seals, in the Antarctic, can swim long distances under the ice. It has now been found that seals kept under a covered surface exhale a large bubble of air which they later return to and reinhale, the carbon dioxide having been dissolved in the water and the oxygen remaining.

Cuttlefishes and Squids

Amongst invertebrate animals there is only one group which can be classed as true swimming animals. These are the cuttlefishes and squids, which belong to the group known as Cephalopoda. This name is derived from two Greek words meaning 'head' and 'foot', and originates from the belief that the tentacles surrounding the head represent part of the 'foot'.

The cephalopods are molluscs, and are allied therefore to oysters, mussels and cockles, though admittedly the external resemblance is not obvious. In many ways they are the most highly evolved of the invertebrate animals and are remarkable for possessing eyes similar in almost all respects to those of the higher animals such as fish and mammals.

Both the cuttlefish and the squid possess ten tentacles armed with suckers. Two of these tentacles are very much longer than the others and can be withdrawn into a cavity near the head. In this respect they differ from their near relative, the octopus, which possesses only eight arms or tentacles, all of the same length.

Within the body is a curious flat shell, which serves to support the animal when it is swimming. In the cuttlefish this shell is hard and limy; it is that white object which we so often find cast up amongst the drift weed on the sea shore, and which we put to such various uses as removing ink stains from our fingers and giving to canaries to sharpen their beaks upon. In the squid, however, the shell is very thin and transparent and is commonly known by the fishermen as a 'pen' on account of its resemblance to an old-fashioned quill pen.

Recent research has shown that the bone of the cuttlefish is organized so that the animal can alter its buoyancy like a submarine. The bone is composed of a number of narrow chambers, and there is specialized tissue which can pump water in or out of these chambers. In the daytime the cuttlefish is heavy and buries in the sand, but when it comes out to hunt at night it increases its buoyancy.

An altogether different type of buoyancy control is shown by certain deep-sea squid. These have large sacs filled with fluid which is lighter than sea water. Their method of control is therefore like that of a bathyscaphe.

These methods are of much interest when contrasted with those of fish, many kinds of which have swim-bladders filled with gas. Much of the weight of a fish is made up of muscle and bone. Many deep-sea fish have achieved decreased density by having weakly developed muscles and skeletons. Other larger fish such as deep-sea sharks, achieve buoyancy by storing oils in their livers and tissues.

Cuttlefish and squid are active swimmers. Usually they move slowly along by a slight wavy motion of the fins, which run along the sides of their bodies; but they can also dart rapidly backwards. To secure the necessary propulsion they draw water into their respiratory cavities and then squirt it out with considerable force through a siphon under the head. Both have within their bodies a little sac containing a black inky powder. This is the sepia that is used in the manufacture of certain brown paints. When alarmed or attacked by an enemy this ink is squirted out into the surrounding water, where it spreads out into a great dark cloud behind which the animal darts rapidly away and so eludes its foe. This method of escape is exactly that used by battleships when they sent out masses of black smoke to form a smoke screen, which drifted out between themselves and the enemy and so hid their movements. One of these animals has been found to create a cloud of ink resembling its own shape which it leaves behind as a decoy.

In appearance the cuttlefish is stumpy compared with the squid, whose body is long and tapering and carries a big triangular fin along its hinder end (Plate 37). In British coastal waters none of the species that occurs reaches much more than 30 cm in body length, but there are some species living out in the deep oceans that are veritable giants of their kind. These monsters, sometimes washed up on Newfoundland coasts, may have bodies some metres in length and tentacles 9 to 12 m long. To add to their horror, each of the powerful suckers ranged along the tentacles is armed with a cruel hook, which can tear the flesh of any prey on to which it fastens. These huge squids, which live in the dimly-lit layers that overlie the cold, black abyss, are very rarely seen, and still more rarely caught. But they can be studied indirectly, for it is on these that the sperm whales feed, and in the stomachs of the whales can be found many remains of these squid, especially the hard, chitinous, beak-like jaws.

Sea Serpents

If the sea serpent really exists surely it can be safely classed as a swimming animal! It is fitting then to conclude this chapter with a few words about this interesting beast.

We can safely say that nearly all the accounts that have been given about sea serpents have been due to mistaken identity. The giant squids mentioned above are probably the main cause of many of the stories that have arisen. These monsters are known at times to come to the surface. What more like a serpent than one of these writhing arms, 9 m in length, coiling one moment on the sea surface, and the next raised aloft far out of the water. In many descriptions the serpent has been said to have been spouting water, an act quite to be expected of a squid.

Other objects that have been suggested to have given the appearance of a serpent are many and varied. Amongst these are a string of porpoises, two basking sharks (it is a common habit for these sharks to swim in pairs one behind the other), long strings of weed, the giant ribbon-fish, and even a flight of birds in single file just above the water.

There are real sea snakes, most poisonous inhabitants of certain tropical seas; but these seldom reach a length of more than 2 or 3 m, and although they are

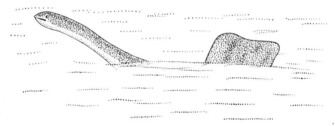

Sketch of an unknown marine animal seen off the coast of Brazil

indeed sea 'serpents', they are not the sea serpents that we all hope it may be our lot to see. There is, however, every possibility that some beast worthy of all that the name implies may still exist in the great depths of the ocean. When we imagine the vast space of the undersea world, stretching for thousands of kilometres north, south, east and west, we realize there are possibilities for the existence of many fearsome monsters; creatures so powerful as to evade all capture. Of the many tales that have been told, most of which can be almost definitely shown to have arisen through mistaken identity, one at least is worthy of mention.

In 1906 two naturalists, Messrs. Meade-Waldo and Nicoll, described in the Proceedings of the Zoological Society an animal that they had seen while cruising on 7 December 1905, in the Earl of Crawford's yacht, the *Valhalla*, off the coast of Brazil. They say: 'At first all that could be seen was a dorsal fin about four feet long, sticking up about two feet from the water; this fin was of a brownish-black colour and much resembled a gigantic piece of ribbon seaweed.' Behind the fin under the water could just be made out the form of a considerable body. 'Suddenly an eel-like neck about six feet long and of the thickness of a man's thigh, having a head shaped like that of a turtle, appeared in front of the fin.' Unfortunately, this curious beast soon disappeared from view and all that may have been seen of it again was on the next night when an animal, which they assert was not a whale, was making such a commotion in the water that it looked as if a submarine was going along just below the surface.

Maybe one day the sea will yield up this, its greatest secret. Maybe not.

It is at any rate certain that the deep waters of the ocean still hold many secrets unknown to man. William Beebe in his deep sea dives in the bathysphere recorded having seen at a depth of 2450 feet (747 m) 'a very large creature at least twenty feet in length' whose identity he was unable to determine. Other giant animals certainly exist; Piccard in his submarine drift beneath the Gulf Stream records seeing a jellyfish 30 feet (9 m) in diameter with tentacles 3 inches (7·6 cm) thick.

A description was given on page 59 of the 'Leptocephalus' stage of the eel. Now it is a remarkable fact that a 'Leptocephalus' which was some 2 m in length has been caught in deep water. If the size of the adult eel of this young stage shows the same relative proportions as does the adult of the common eel then it must indeed be a veritable sea serpent! It should not be forgotten also that it is not long since the 'fossil' fish, the cœlocanth *Latimeria*, was caught alive, when it had been thought that it had been extinct for millions of years.

Chapter 5

DRIFTING LIFE

We have dealt so far with those animals and plants which live either on the sea bottom, or swim actively through the water above the bottom. There is yet another community of organisms in the sea whose existence is, perhaps, not generally realized. It consists of countless numbers of animals and plants which drift about in the water layers at all depths between the sea surface and the bottom, at the mercy of tide and current. They are nearly all small, most of them minute, and many only visible under the high-powered microscope.

It is extremely easy to demonstrate their presence; it is merely necessary to drag through the sea for a few minutes a small cone-shaped net, or 'tow-net', made of fine muslin, nylon or silk; or, if one is on an ocean liner, it is even possible to obtain them by hanging a small muslin bag under the salt water tap in the bathroom and just letting the water run.

At times this drifting life is so abundant that it colours the sea for many kilometres around. Such expressions as 'red water', 'yellow water' and 'green water' are used by fishermen, who become expert at deciding where to shoot their mackerel or pilchard nets by slight differences in the tint of the water that landsmen would not notice. In every case the characteristic tinge given to the sea is known to be due to the presence of certain kinds of organisms in innumerable quantities. Darwin in his *Voyage of a Naturalist* noted such discoloration, and mentions passing through two patches of reddish-coloured water, 'one of which must have extended over several square miles'. 'What incalculable numbers of these microscopical animals!' he exclaims. 'The colour of the water, as seen at some distance, was like that of a river which has flowed through a red clay district; but under the shade of the vessel's side it was quite as dark as chocolate. The line where the red and blue water joined was distinctly defined.'

'Stinking water' is another expression used by fishermen, and in this case a characteristic odour is given to the sea also by the presence of hordes of certain organisms.

A very large number of these small drifting creatures are capable of emitting a phosphorescent light. It is these that make the sea sparkle with little glowing points of fire when we dip our oars into the water on a dark night. Occasionally, some luminescing animals will swarm together in such countless numbers that on calm still nights the whole sea surface seems to glow with a pale cold light.

Darwin again describes such a sight in picturesque terms. He says, 'While sailing a little south of the Plata on one very dark night, the sea presented a wonderful and most beautiful spectacle. There was a fresh breeze, and every part of the surface, which during the day is seen as foam, now glowed with a pale light. The vessel drove before her bows two billows of liquid phosphorus, and in her wake she was followed by a milky train. As far as the eye reached the crest of every wave was bright, and the sky above the horizon, from the reflected flare of these livid flames, was not so utterly obscure as over the vault of the heavens.'

Over eighty years ago the word 'plankton'* was used by a German professor to embrace all this drifting life, and the word is now in general use among those interested in the science of the sea.

Plankton Plants

One naturally wishes to know what kinds of creatures these are that make up the almost infinite multitudes that drift freely throughout the water layers. The plankton is now known to play a part of fundamental importance in the economy of the sea, and in this respect the organisms that deserve our first consideration are the microscopic plants. These are not like the plants on land, but consist generally each only of a single cell; nevertheless, they are plants in the true sense of the word, because each contains within its cell colouring matter closely similar to that so characteristic of our land vegetation. These little plants are known as 'diatoms'; they are so called because they have, surrounding their cell-walls, glass-like protective shells composed of two halves—two lid-like structures that fit one into the other, and thus enclose the body of the plant in a little box.

In order to catch these diatoms it is necessary to use a net made of the finest muslin or silk, as they are, mostly, so small that they will pass through the meshes of ordinary coarse muslin. The catch, to the naked eye, will look like a greenish-brown scum, and if placed under the microscope will be seen to consist of a jumble of interlacing spines, amongst which may be noticed green oblongs, squares and circles. To find out the true nature of the catch it must be diluted with sea water and only a drop examined, when it will be found that the diatoms are much fewer in number and separated one from the other so that their true structures can be made out. Some will be like little circular discs, others oblong with little spines or horns projecting from each corner, and others strung together to form chains of tiny boxes covered with delicate, interlacing, hair-like projections. On Plate 32 are given drawings of some of the commonest diatoms that would be found in any catch from the northern and more temperate regions of the Atlantic Ocean and the seas around its border. There are many thousands of different species occurring in the world, but this is no place to confuse the reader with a medley of Latin names. For those who may take a deeper interest in making the acquaintance of the different kinds of diatoms and who find delight in observing the marvellously delicate and beautifully designed structures of these minute plants through the microscope, a list of books will be found at the end of this volume.

In addition to the diatoms there are other single-celled organisms that help to swell the plant life of the drifting community. They are especially remarkable

* Gk. πλαγκτός, wandering.

because, besides containing the colouring-matter of plants, they possess two tiny structures like the lashes of a whip which by vigorous waving motion serve as a means of propelling the creature through the water. Now, a true plant derives all its nourishment from gases and dissolved salts that are absorbed through the cell walls; it never takes in solid particles of food as an animal does. Many of these little creatures, known as dinoflagellates, have no coloration and are able to swallow solid particles of food through a small depression on their cell surface. There are, however, a few which possess colouring matter and also swallow solid particles; because they are able to feed like plants, and at the same time utilize solid food like animals, they are a continual source of bickering in scientific circles, the botanists claiming them as plants and the zoologists maintaining that they are animals!

Many of these flagellates consist only of a little naked cell with its whip-like lashes and are, in consequence, extremely delicate and easily destroyed by the net; others, however, possess wonderfully designed skeletons made up of little plates which may carry spines and wings, giving the plants a beautiful appearance (Plate 9). Some dinoflagellates are the cause of the coloured water mentioned above. The *Gonyaulax* figured in Plate 9 has been known to occur in such profusion as to colour the water red, and other species are at times so abundant that when they die and decay they cause the death of great numbers of fish. Other species are known to be poisonous and when ingested by shellfish such as mussels cause poisoning in human beings

There is yet another group of plankton plants known as coccolithophores. These also are unicellular but are characterized by the presence of numerous calcareous plates embedded in the cell, the different shapes of which serve as a means of identification of the numerous species.

These three groups, the diatoms, the dinoflagellates, and the coccolithophores, are the most important constituents of the drifting plant life. The diatoms are most abundant in the colder waters of the temperate and polar seas, while the two others are more characteristic of the warm waters of tropical and sub-tropical regions.

These plants are at times extremely abundant; they have often been reported to be so thick in the Baltic that a thimbleful of water would contain more than a thousand individuals. Were it not for these countless myriads of small plants we can safely say that the great oceans and seas of the world would be valueless to us as reservoirs from which to draw much of our food in the form of fish and other edible marine animals, for the small plants of the plankton are indeed the pasturage of the sea. They form the food of millions of small animals living in the drifting community on which larger animals prey. On land, man and beast alike are ultimately dependent on the grass and herbs of the field for food, the animals which feed on the grass being eaten by man; so also in the sea, the ultimate food supply is to be found in the drifting microscopic plants. But whereas on land the plants are substantial and can be directly eaten by large animals, in the sea they are minute and are first eaten by the small drifting animals, which are in turn swallowed by larger creatures, and so on until, eventually, the fish forms food for man. In fact certain chemical constituents of our food have been traced to these tiny plants. To most the word 'vitamin' is well known. It is the name given to certain chemical bodies, which, although present in minute quantities only in our food, are essential to our health and well-being. Their absence gives rise to such diseases as scurvy and rickets. Cod-liver oil contains vitamins A and D which are partly responsible for its great medicinal value. The presence of these vitamins has been traced from the liver of the cod to the insides of the capelin (*Mallotus villosus*), a little fish that forms a large portion of the cod's food, and the swarms of which bring the cod together in vast shoals on the Newfoundland banks. From the capelin vitamin A has been traced to the minute animals on which it feeds. It is in the little plants—the diatoms—which nourish the animal plankton—that the precursors of the vitamins are presumably made.

Collectively this plant community is known as phytoplankton in contrast to the zooplankton now to be described.

Plankton Animals

We have mentioned above that the drifting plants form the chief food of the small animals of the same community. Of what is this animal population chiefly composed ? Actually almost every group of the invertebrate animal kingdom has representatives in the plankton. In addition, the young of many kinds of fish live for a shorter or longer period a free, drifting existence. But of far the greatest importance in the animal plankton is a group of crustacean organisms known as 'copepods' or 'oar-feet'. They are all small, the largest being under 13 mm in length. While there are very many species included in the group of copepods there is one that stands out far before all others in numbers and in importance in the chain of food organisms that links the fishes with the drifting plants. This little animal is unfortunately unknown to most people, because it can only be caught with the aid of a tow-net, and when captured is so small that it is regarded as insignificant. But if ever an animal merited attention it is the small copepod which goes by the Latin name of *Calanus finmarchicus* (Plate 22). It is unfortunate that it has no real popular name of its own; but we shall call it in these pages '*Calanus*' for short, in the hope that some day *Calanus* will be just such a common everyday expression as shrimp, crab, or prawn.

Calanus is an inhabitant of the cold northern waters, where it forms one of the chief items of food of that most important of all food fish, the herring, and is even sufficiently abundant to aid in the building up of the enormous bodies of two of the Atlantic species of whales, one of which has been described as having been seen 'in still weather, skimming on the surface of the water to take in a sort of reddish spawn or brett, as some call it, that at times will lie on top of the water for a mile together.' The spawn or brett is, of course, the little copepod, *Calanus*, which occurs at times in such swarms as to colour the water, whence it has acquired the name from fishermen of 'red feed'. It will give you some idea of the numbers of these little creatures if two examples are given of exceptionally heavy catches. In the Gulf of Maine, by towing a conical-shaped net with a circular opening of 1 m diameter behind the boat for fifteen minutes, over

2,500,000 specimens of *Calanus* have been caught, or enough to fill ten ½-litre tumblers solid. Again, near Iceland, 200,000 have been recorded from a five minutes' tow.

There are many other species of copepods but none to compare with *Calanus* in importance, although many far excel it in beauty, especially some of those that come from warmer and more tropical climes. Some of these are equipped with the most beautiful array of feathery spines; others are iridescent and shine with all the colours of the rainbow (Plate 21).

But next in importance as food for other marine animals, if not perhaps the most important, are shrimp-like animals that, like *Calanus*, are rarely seen and almost completely unknown to the world in general. These are known as euphausiids (Plate 22), or 'krill' by the Norwegians. They are about 38 mm in length, but are so abundant that they form a large part of the food of many of the northern fishes, and are the chief food of nearly all of the whalebone whales. They may form swarms so dense that there are over 500,000 individually in a cubic-metre. Their bodies are quite transparent except for the presence of minute red spots, and they possess enormous black eyes and on this account the fishermen of the west coasts of Scotland call them 'Suil dhu' or 'black eye'. But they are most remarkable because along the sides of their bodies are numerous little organs that can blaze up into brilliant luminescence at will. More will be said about this luminescence under Chapter 8, but suffice it to say that it has been recorded that with six of these little animals in a jar of water, flashing on their lights, it is just possible to read newspaper print!

There are, besides, many microscopic, single-celled animals dwelling in the drifting community. Of these, perhaps, the group of animals known as radiolarians are of the greatest interest. These little unicellular creatures are noteworthy for possessing silicious skeletons on which their protoplasm is supported. Although these skeletons are so minute, they are fashioned in the most beautiful and symmetrical patterns. Almost every conceivable shape is to be found among them.

Another unicellular animal that is very commonly found is the *Globigerina*, which builds a calcareous shell made up of a number of connecting compartments.

So abundant are these two groups, the radiolarians and the globigerinas, in certain parts of the ocean, that when they die their skeletons, sinking to the bottom, form characteristic deposits. These are the radiolarian and globigerina oozes mentioned in Chapter 3; in certain localities also diatom oozes are formed from the rain of siliceous frustules, the skeletons of the dead diatoms.

Enough has been said of the most important members of the animal plankton. Let us now consider some of the more grotesque and unusual forms. Ordinary shellfish or molluscs are heavy lumbering creatures; yet there are some members of this group of animals that are delicate enough to drift about in the water layers amongst the plankton community without fear of sinking rapidly to the bottom. They are the sea butterflies, a most fascinating group of marine organisms. Perfectly transparent, some carrying a delicate paper-like shell, they move through the water by the rapid flapping of what appear to be wings. These wings are in reality modifications of the foot, that solid muscular organ on which most molluscs creep. These animals are also so numerous in some parts of the ocean that their empty shells form deposits on the sea floor, the pteropod ooze (see p. 37).

It is a characteristic of nearly all the plankton animals that live in the upper layers of the sea, that they are almost transparent. If one looks at a tow-net catch

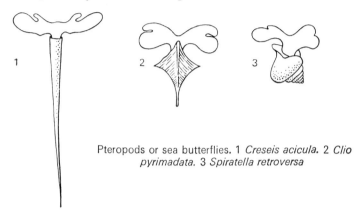

Pteropods or sea butterflies. 1 *Creseis acicula*. 2 *Clio pyrimadata*. 3 *Spiratella retroversa*

that has been placed in a glass jar full of sea water, it is at first very hard to see the various animals on account of their transparency. A very common creature in the catch is the arrow worm, or *Sagitta*; this is thought to be a relative of the true worms, like the rag-worm, although it is very unlike them in appearance. It looks just like a little glass rod about 10–20 mm in length; but for all its

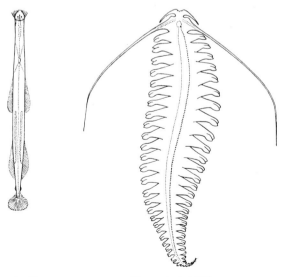

Sagitta (×2)   *Tomopteris* (×2)

apparent delicacy it is a very voracious creature. Surrounding its mouth are a number of powerful hook-like teeth, and with these it can seize on its prey, which it rapidly devours. Often it can be seen to have within its stomach one or two of the copepod *Calanus*, and, when very young herring are abundant, it will capture them and eat them even though they be as long as itself.

Another very beautiful planktonic worm is *Tomopteris*, which has a row of wing-like feet down either side of its body, with which it paddles its way through the water with a curious wriggling motion.

In warm ocean waters there are commonly to be found numbers of animals known as salps. These are closely allied to the common sea squirts, which lead a sedentary existence fixed to rocks and piers; but unlike its relative, the salp lives a free drifting existence in the open waters. They are 25 mm or more in length, shaped like a barrel, and perfectly transparent. They are not usually to be

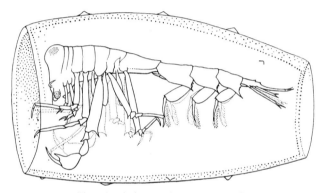

Phronima in barrel-shaped 'house' (× 4)

seen in northern Atlantic waters, and when found are a fairly reliable indication of the presence of Gulf Stream water.

Belonging to the same group of animals is *Pyrosoma*, so remarkable for its luminescence. This animal is reputed to suffer a curious indignity at the hands of a small crustacean, the *Phronima*, which is said to eat off all the living portion of the *Pyrosoma* (which is really a colony of animals) and then retire inside the barrel-like skin that is left. There is some doubt as to what this house is really made of, but whatever its origin *Phronima* are often caught in oceanic collections living in their gelatinous barrel-shaped homes, surrounded by a brood of young.

But the members of the plankton are too numerous to mention here. The best book is nature itself, and therefore at the first opportunity one should make a small tow-net, take a small boat and row out one or two kilometres from the shore on a calm day, drop the net overboard until it is a metre or so below the surface, and then tow it slowly along for a few minutes. The catch will be sufficient reward for the trouble, and the sight of the delicate creatures will open one's eyes to the amazing abundance and diversity of this drifting life that peoples the sea in every region of the globe.

The few animals that have so far been mentioned constitute part of what is known as the 'permanent plankton'; that is, they live as drifting organisms for the whole of their lives. There are yet other members of the plankton which have adopted this mode of life for only a short period in their life-history. In the coastal regions nearly every animal we find has at some time drifted freely and aimlessly about in the water layers above the bottom. The young of all the smaller marine animals are, when first they hatch from the egg, extremely small, and in consequence have insufficient swimming power to cover large distances, and are unable to cope with the tides and currents. At this stage, they rise up from the bottom, if hatched there, and become members of the plankton. Here they will remain for shorter or longer periods, growing and developing until such time as they have assumed their adult characters or are large and strong enough to seek for themselves their natural home on the bottom.

Worms, starfish, crabs, lobsters, oysters, sea squirts and even the majority of fishes in the sea, spend the first days of their lives drifting about in this manner (see Plate 33). One of the main advantages of this mode of life is that it ensures complete dispersal of the young. Many marine animals, such as hydroids and mussels, live fixed to the rocks and bottom or are at most very sluggish and do not move far afield. It is obvious, therefore, that if their offspring were born and hatched 'at home' the parental abode would soon become so crowded with young and half-grown individuals that there would be no room to turn. But if the latter are sent out to fend for themselves in the upper water layers they will be carried far and wide by the currents, and those that survive will settle to the bottom many kilometres from where their parents lived.

These drifting young of many marine animals are very different in appearance from their parents; indeed, it is safe to say that if they were shown to a novice it would be impossible for him to say into what they would grow. Because of these extraordinary differences between adults and young, the early stages of many marine animals, when first discovered by naturalists in the plankton catches, were regarded as new species of animals and described and given names of their own. It was only by rearing them, or piecing together successive stages in the life-histories from the catches, that the adult into which they were going to develop could be discovered. Who, for instance, without knowing beforehand, would dream of suggesting that the little animal figured in Plate 33 was the young of the common edible crab? When first discovered it was not recognized as such and was given the name 'Zoea'; on this account it is now known as the zoea stage of the crab. After hatching, the young crab remains in this stage for some time during which it undergoes several moults. Finally, it suddenly moults into a stage quite unlike the zoea and more like the adult crab; albeit when first found it was not recognized as such and was labelled 'Megalopa' (Plate 33). From this stage it moults into a perfect little crab. So we see that in the life-history of one animal, the crab, the successive stages are so different that no fewer than three animals have been thought to exist, the zoea, the megalopa and the crab, which were in reality all crabs.

These animals, which appear for only a short period in the drifting community form the temporary members of the plankton, and their appearance in the upper water layers depends, of course, on the times at which the parents spawn, and this gives rise to the seasonal changes in the plankton mentioned on page 173. It is chiefly in the coastal regions that these temporary drifters are found because of the great wealth of bottom life in the shallower sub-littoral zone.

Distribution of Plankton

Having given the reader some idea of what composes this drifting life or plankton, let us now examine what is known of the distribution of the plants and animals that form it in the sea.

To begin with it must be realized that besides the immense horizontal area of all the oceans and seas of the world, all of which are inhabitable, the waters also have a considerable depth, the average of which is as much as 3 km. Seeing then that organisms that live in the water layers themselves are free to move up and down, as well as in a horizontal direction, it is evident that we have two types of distribution to deal with, namely, horizontal or geographical distribution, and vertical or depth distribution.

Much work has been done on the geographical distribution of the animals and plants of the plankton by large oceanographical expeditions which have been sent out by different countries during the last hundred years, and it is fairly established that all the waters of the globe are inhabited by them. Just as on land, the various species of plants and animals have their own distribution, some living in tropical climes and others in the far north or south, so it is noticed that many members of the plankton are to be found only in certain regions and may give the catches made in those localities a characteristic appearance. Amongst diatoms, for instance, there are species that are normally found only in coastal regions, others that occur only in the open ocean waters of tropical and sub-tropical regions, and others again that characterize the catches of northern waters. But while species of these little drifting plants are to be found almost anywhere in the oceans all over the world, at certain times of the year the cool temperate and the cold arctic and antarctic waters carry this diatom life in quantities far exceeding the other regions. The waters that bathe the shores of the British Isles, of Holland, Sweden, Denmark, Norway and Iceland, on the east, and of Greenland, Newfoundland and the Gulf of Maine in America on the west, possess a richer spring and summer pasturage of microscopic plant life than is to be found in any other locality in the North Atlantic Ocean. The significance of the plankton plants in the general economy of the sea will then become at once apparent, when it is realized that it is precisely these regions, the North Sea, the Baltic, the Norwegian and Greenland Seas and the banks of Newfoundland, that give rise to the greatest fisheries in the Atlantic.

Dependent on the diatoms are, of course, the small animals of the plankton, and it is natural to suppose that where the plant life is most abundant there will be found the greatest quantities of animal life. Such indeed is the case. The abundance of that little creature, *Calanus*, and of the shrimp-like euphausiids, has already been dwelt upon and they too are to be found chiefly in the regions outlined above.

But, while the plankton is present in greatest quantities in these northern waters, it is remarkable that compared with the plankton of the warm and tropical regions the numbers of different kinds of animals are extremely few. The catches made in the northern waters can almost be described as monotonous in composition; that is, although they are so large, they will be made up of only comparatively few species of animals. To the collector then the catches made in warmer regions prove vastly more interesting on account of the wealth of different species to be found there, even though each be present only in small numbers. One example is sufficient to show how marked this difference is. Of roughly five hundred species of copepods, 87% are to be found in the warm regions, while only 3% are found in the cold northern region and only 6% in the southern. This phenomenon holds good for all the different groups of animals represented in the plankton.

Apart from the excessive abundance of drifting life in the colder waters, it is to be noticed that everywhere the coastal regions are considerably richer than the waters of the open ocean. The central regions of the North and South Atlantic Oceans are the poorest in plankton life; that is the areas lying a little north and a little south of the equator. Actually in equatorial regions the plankton is a little richer. The deep blue Sargasso Sea, in the centre of the North Atlantic, is probably as barren as any region in the world. There are several factors that together are responsible for the abundance or poverty of plankton life, but of greatest importance is the presence or absence of certain nutrient salts, dissolved in the water on which the drifting plants depend for their growth. In localities where these bodies are richest the plants will be most abundant, and, consequently, the animals which depend on them for their food supply. A discussion giving the basic principles that underlie this plankton production will be found in Chapter 12; it would be out of place to enlarge on the problem here until the reader has made himself acquainted with some of the properties of sea water that are outlined in Chapter 11.

In Plate 24 is given a chart of the North Atlantic Ocean and surrounding seas, in which the density of the standing crop of plankton life (see p. 177) is shown diagrammatically. The richest areas are shown by the green coloration, the regions of greatest density of life being the deepest green. The green colour gradually shades off into blue which can be taken to represent a poverty of plankton organisms.

Having dealt with the geographical distribution of the plankton, let us turn now to consider at what depths this drifting life is to be found and where it is most abundant. Taking the plants first, consisting of the diatoms and the dinoflagellates, collections made by research vessels from different depths show that it is only the upper water layers of the sea, from the surface down to about 200 m, that contain drifting plant life in any quantity.

A moment's thought will satisfy anyone that this must of necessity be the case, since all plants are dependent on the sun's light for their life and the deeper we go into the water the less light is there present. In fact it has been shown that at little more than a depth of about 20 m in the English Channel off Plymouth the amount of light present is already similar to that in the heart of an

G

English wood. In the clearer open waters of the ocean the light can perforce penetrate deeper, but it is certain that at a depth of 200 m it is already so dim that few
plants can live in a healthy condition. More will be found on the subject of the
light beneath the sea surface in Chapter 11. Although there is a depth limit at
which the normal life of the plant becomes impossible, yet at times collections
may show their presence at still greater depths, the likelihood is however that
these plants will be in a dying condition and will have sunk to those depths
under their own weight.

Now animals, as we know, are not so dependent on light directly; in fact there
are many animals on land that seem to prefer darkness and are nocturnal in
their habits, shunning the light of day.

The same applies to the sea animals. There are many that live in the dark
deep layers; recent research has shown that there is no depth at which some
plankton organisms do not exist.

All the animals are not, however, found evenly distributed in the water layers
from top to bottom. Each animal seems to show a definite preference for some
particular depth region; for instance, there are some that live always in the
upper hundred metres or so, rarely penetrating to deeper levels; others again will
never be caught between two hundred metres and the surface, but always live
deeper. On account of these differences in depth distribution, exhibited by the
various animals of the plankton, we find that catches made from different levels
are each quite characteristic and distinct one from another in their composition,
in the same way that collections from different geographical regions are each
characterized by the presence of those organisms that are prevalent in the locality
in which the catch was made.

It is a general fact that in the daytime the animals that live in the layers quite
near the surface are very few. This is especially the case in the summer months.
For instance, in coastal waters it is necessary in the daytime to fish at a depth of
20 to 30 m in order to get the largest catches. In the open ocean the depths at
which the greatest assemblages of plankton animals are found are generally
considerably deeper. Now, as has been said, this is the case in the daytime; but
it is a remarkable fact that at night matters are quite different, and a curious
change comes over the plankton distribution. On the approach of dusk, just
after the sun has set, all the animals begin to swim in an upward direction, so
that by about nine or ten o'clock, in the summer months, the surface layers, so
barren in the daytime, have been enriched by animals that have swum up from
the deeper levels. If the results of collecting at different depths in daylight and
at dusk are compared, it can be clearly seen that whereas in the daytime the
layers down to about 10 m are very poor in plankton compared with the deeper
levels, at dusk the upper collections are quite as large as the deeper ones.
It will be noticed that the bottom catch at dusk is also greater than that taken
in the daytime. This is because of the addition of large quantities of animals
that have come up from deeper levels quite near the sea bottom; in fact, there
are certain small creatures that live on the bottom itself in the daytime and
which move upwards towards the surface at night; they may never actually
have time to reach the surface itself, because at dawn all the animals begin to

move downwards again to take up their abodes at their usual day levels (Plate 25). Herring fishermen when using drift nets always fish at night, for the herring, like plankton animals, also move upwards at night.

This phenomenon of upward movement at night, or vertical migration, throws considerable light on why the different animals prefer certain depths in the daytime.

It seems almost certain that the causes of the up and down movements are the changes in the strength of light experienced by the animals; this is further confirmed by the fact that many animals live deeper in the bright sunny days of midsummer than they do earlier in the year, when the light entering the water is not so strong. From this, then, we see that most of the plankton animals tend to avoid strong light and prefer the dimly lit conditions of the deeper layers; at the same time all the evidence goes to show that each animal shows a preference for a certain strength of light to which it is adapted. Towards the evening they follow the 'optimum' strength of light towards the surface as night draws on, but in the dark there is no light stimulus and they are free to move anywhere (see Plate 25); at dawn they once more pick up their optimum intensity and move downwards as the daylight strengthens. It is also noticeable that the older an animal becomes the more it shuns the light, and it is generally the younger stages that are found nearer the surface. This is not however, always so; the early stages of *Velella* or 'by the wind sailor' are found at very deep levels, while the adult, as mentioned on page 90, floats right on the surface of the water, where it is blown hither and thither by the winds.

Adaptations for Suspension

In considering this drifting life it may have struck the reader that it is a curious thing that all these organisms, many of which, such as the diatoms, are practically incapable of any independent movements, should remain suspended in the water. Why do they not sink rapidly to the bottom of the sea? The answer to this question rests in the demonstration of some of the remarkable structural adaptations, which fit most of the plankton organisms to the conditions under which they live.

Their main requirement is that by some means or other they should be able to maintain the level at which they are drifting. The means by which this capacity is attained are many and varied, the general aim being to reduce the organism's specific gravity directly until it is the same as, or less than, that of the surrounding sea water, or to obtain a similar effect in a more indirect manner.

Some species have achieved the capacity of becoming lighter than water. The most obvious examples occur in a group of jellyfishes known as siphonophores. The name of the Portuguese man-o'-war (Plate 13), a stinging jellyfish, is well known to all. This animal possesses a specially designed 'float' into which gas is secreted by a specialized gland. This gas-filled reservoir projects above the surface of the sea and acts as a sail, by means of which the wind blows the jellyfish along, transporting it from place to place with tentacles extended below

the surface ready to seize any unwary prey that they may touch, instantly paralys-
ing it with their batteries of stinging cells. Another closely allied form is the
Velella, mentioned above. This likewise has a small gas-filled sail. The animal,
when seen alive, is of a fairy-like delicacy, possessing this transparent, papery
sail, situated above the centre of the body: on the under surface is the 'mouth',
centrally placed and surrounded by delicate mobile tentacles of a sky-blue tint.
These little creatures, which reach a size of 25 to 50 mm in length, are natives of
the warmer ocean waters and the Mediterranean. However, they are occasionally
to be found stranded along the western and south-western shores of the British
Isles after prolonged southerly winds, that have wafted them speedily over the
surface of the sea from the warmer latitudes.

In order to reduce the specific gravity as nearly as possible to that of sea water,
many animals make use of fats and oils formed in their bodies. These oils
being lighter than water, tend to diminish the weight of the animals that contain
them.

By varying the composition and water content of their body fluids most
plankton animals and plants are probably able to adjust their density. But in
addition to these direct methods of obtaining buoyancy it is possible that another
striking characteristic of many plankton organisms may play its part, and it is
a feature which certainly adds much interest for the observer. If we drop a
stone into water, and watch it sinking, we shall notice that its sinking speed is
considerably less than if it were falling through air only. The speed is reduced
by the frictional resistance set up by the stone as it moves through the water.
This property of resistance to the movement of a body is known as viscosity.
Some liquids are naturally more viscous than others; they are said to have a high
viscosity. Treacle is a very viscous fluid; many oils also are highly viscous. Thus
the frictional resistance to a falling body would be greater in treacle than in sea
water. Now, of course, the total frictional resistance experienced by a falling
body depends on the amount of area exposed against the fluid through which it
is moving. We know that a slate takes longer to sink flatways than if it is on edge.
Clearly then, if the frictional resistance can be made infinitely great compared
with the actual weight of the body itself, a stage will be reached at which it
would counteract the force due to gravity and the body would no longer sink,
but become suspended in the water.

The structure of many plankton organisms appears to show an attempt to
increase the frictional resistance, which may help towards keeping the organism
suspended in the water.

The effect is brought about either by greatly increasing the surface area by
means of long spines or feather-like projections, or by extreme flattening so
that the creature is like a leaf and when lying horizontally in the water is prevented from sinking except at a very low speed. The spiny and hair-like processes on many of the plankton diatoms depicted in (Plates 22 and 32) may serve
this purpose, although some may also help directional movement. The diatoms
themselves are so very minute that the many spines must increase their surface
area comparatively to an enormous degree, and they become literally suspended
in the water, so slow is their sinking speed. The rate at which a diatom will sink

through the water has been measured, and, of course, varies for the different species according to their shape. Let us take as an example a species of *Chæto-ceros* (Plate 32); this takes on an average about $4\frac{3}{4}$ hours to sink 1 m, or 3 minutes 12 seconds a centimetre. This was in still water; but in nature, near the sea surface, owing to wave action there are continual little swirls and eddies which will bring it up or down at a much faster rate than it sinks.

Now, the animals, of course, on account of their larger size and weight will sink considerably faster than plants, but they also possess the power of loco-motion, by which they can regain their level. The presence of spines and feathery processes on the animals' bodies, must slow down the rate at which they sink, and save them from having to be so continually on the move to keep up in the water. If a living plankton animal is watched it will be seen to make a rapid up-ward movement by swimming and will then rest while it sinks only slowly through the distance that it so rapidly moved up through.

An extreme example of flattening in an animal is to be found in the case of the

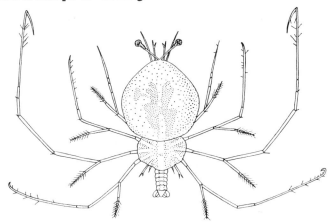

Larva of rock lobster, *Palinurus elephas.* (× *ca*. 6) (After Russell)

larva of the common crawfish or rock lobster. This is, of course, a different species from the lobster that is most eaten for food in the British Isles. It is the langouste, which is, however, much preferred as a delicacy by the French.

The larva is a most curious sight. The body, measuring about 6 mm across in the oldest individuals, is quite transparent and flattened like a piece of paper, while the long feathery legs and projecting eyes stick out all round and must help in preventing the animal from sinking through the water.

It is not necessary to resort to experiment to prove that these structures may be aids to suspension in the water. It is known that the viscosity of water varies with the temperature, a rise in temperature lowering the viscosity to a marked degree, so that the resistance the water offers to a body sinking in it is consider-ably lessened. If, then, an animal or plant lives normally in warm water it will need to be extra well equipped with hairs and spines compared with its cousins who live in cold waters.

If organisms from cold and temperate regions are compared with those from the tropics, it is indeed found that the latter exhibit to a much greater degree these structural excrescences, making many of them bizarre and grotesque in appearance. It has even been found that the same species of dinoflagellate may differ from place to place in form. For instance, in two individuals of the same species, the one from tropical seas has very much longer horns than those of its relative from the colder waters around Britain.

Applications of Research

In these few pages an attempt has been made to put before the reader some of the main features of that remarkable community of drifting life, the plankton. Its importance cannot be over-estimated and plankton studies form a considerable part of modern investigations into the resources of the sea. This chapter has been devoted mainly to describing what the plankton is and where and how it lives, but in Chapter 12 further information will be given of the part played by these small organisms in the light of current research.

We cannot conclude, however, without some mention of applications of plankton research to fishery investigations. Herring, for instance, feed entirely on plankton animals. Their distribution is governed much of the time by availability of food. Often, therefore, the presence of certain plankton animals such as *Calanus* in great quantities is a fair indication of the likelihood of shoals of herring in the neighbourhood. The finding of areas for tuna fishing in the open oceans has likewise followed the discovery of those regions in which plankton is unusually rich.

The study of plankton organisms also proves of use in helping to show the movements of water masses. Some species of plankton animals are restricted to waters of certain types. The presence of oceanic water can, for instance, be detected in coastal areas by the presence of certain oceanic plankton organisms which therefore give an additional indication of the origin of the water to those chemical and physical characteristics usually made use of by hydrographers. Such organisms are known as plankton indicators and they are sensitive to minute changes which cannot yet be distinguished by chemical analysis of the water.

The study of this interesting drifting community thus assumes practical importance and a special continuous plankton-collecting apparatus, designed by Sir Alister Hardy, is now available for towing behind ships running on regular commercial routes. This instrument stores and preserves the plankton on long rolls of silk for later examination in the laboratory. As a result charts can now be produced each year showing the changing distribution of the plankton over extensive areas of the seas and oceans.

Chapter 6
BORING LIFE

Of all the creatures which inhabit the sea, few are more interesting than those which bore into wood or stone and live within the burrows they construct. None certainly does so much damage as the wood borers, notably the dreaded ship-worm which, since the dawn of history, has been the cause of grave damage to wooden ships. The galleys of Greece and Rome in classic times, those of Venice in the Middle Ages, and Drake's famous *Golden Hind*, all were riddled with the burrows of the shipworm which has even threatened, by its attack on the dykes, the very existence of Holland. Although in modern times the steel hulls of the majority of ships have nothing to fear from it, the shipworm still does great dam-age to wharves and piers made of wood; so great indeed that in Great Britain, in the U.S.A. and elsewhere extensive investigations have been conducted to discover some means of combating its ravages and those of its accomplices.

Wood Borers

The wood borers may be divided into two groups, molluscan and crustacean. The former is the more important including, as it does, the shipworms. Despite the common name and the naked, worm-like body, these are really bivalve molluscs adapted for an unexpected mode of life and which have, in conse-quence, become very unlike their relatives such as cockles or mussels. When taken out of its burrow, the shipworm, as shown overleaf, is seen to consist of a long, naked body with at one end a pair of small, peculiarly shaped shell valves, and at the other a pair of delicate tubes or siphons. The former is the front end of the animal. It lies at the inner end of the burrow and bears the boring organs and the mouth. The siphons are the only parts of the animal which project from the burrow; they are instantly withdrawn when they are touched and the opening of the boring, dumb-bell shaped and lined with lime, is closed by the pushing out-ward of a pair of shelly pallets, club-shaped in some species but segmented in others, as in the figure. These are unique to shipworms; they can be moved in and out because the animal is actually attached to the boring at the base of the siphons. Within the body the organs are greatly extended. The principal organs are near the front end, but along almost the entire length there stretches a cavity divided horizontally by a delicate lattice-work of the ciliated gills (see p. 230). Water enters the body of the animal through the under or inhalant siphon and is then filtered through the lattice-work, leaving behind it any food particles which are carried to the mouth, then passing into the upper chamber from which it is

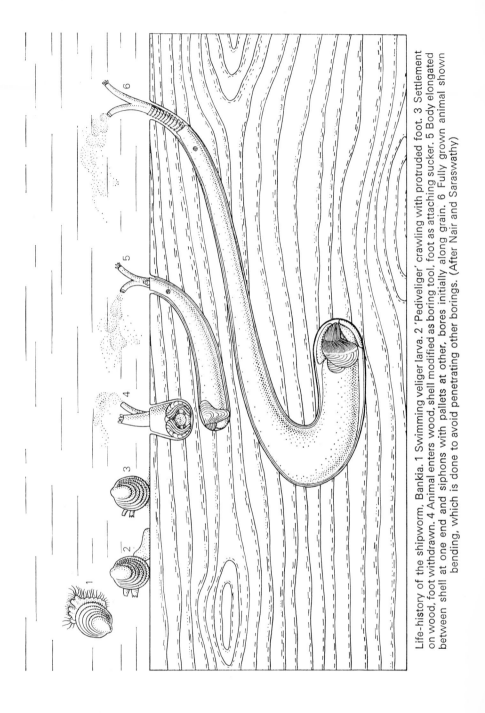

Life-history of the shipworm, Bankia. 1 Swimming veliger larva. 2 'Pediveliger' crawling with protruded foot. 3 Settlement on wood, foot withdrawn. 4 Animal enters wood, shell modified as boring tool, foot as attaching sucker. 5 Body elongated between shell at one end and siphons with pallets at other, bores initially along grain. 6 Fully grown animal shown bending, which is done to avoid penetrating other borings. (After Nair and Saraswathy)

expelled by way of the exhalant siphon. In this way the animal obtains both oxygen for respiration and a certain amount of food.

Now let us consider how it bores its way through the wood. To understand this it is necessary to refer to the figure opposite which shows the position of the boring organs while they are in operation. Although many theories have been advanced on the subject, it is now certain that boring is carried out by the action

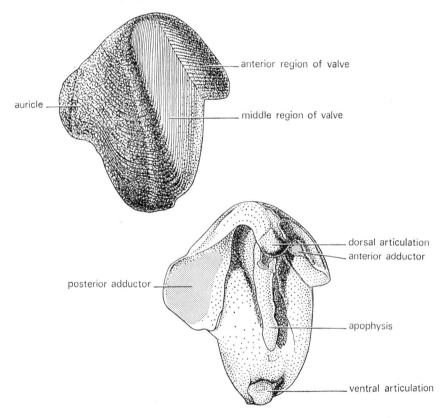

anterior region of valve

auricle

middle region of valve

dorsal articulation

anterior adductor

posterior adductor

apophysis

ventral articulation

Shell of a shipworm, *Teredo*. Above, outer view of right valve. Below, inner view of left valve

of the small shell valves, which are specially adapted for this purpose. As shown here they are globular and can be divided into three regions, a hinder portion in the form of a broad wing and known as the auricle, a middle portion which forms the major part of the shell, and a portion in front of this which extends for less than half the depth of the middle portion and is then cut away sharply at right angles. The surfaces of these two latter regions are covered with sharply pointed ridges, those on the former passing diagonally across the anterior third of its surface, while those on the latter run parallel with the sharply cut lower margin. Transferring our attention to the inner surface of the shell, we see that there are two knobs at top and bottom of the middle region, while from the upper of these there projects down a long process known as the apophysis. The two halves of the shell are in contact by way of the two knobs so that the valves are able to rock backwards and forwards upon these two points, first the hinder and then the frontal regions coming closer to one another.

This rocking process takes place regularly in life, the motive power being supplied by the two adductor muscles, one of which runs between the hinder regions of the shell and the other between the frontal regions. From between the valves in front there projects a round sucker, representing a modification of the foot or organ of movement in such bivalves as the cockle. The muscles which work the foot are attached to the apophysis and by their contractions enable it to grip the wood at the head of the burrow. At the same time a fold of tissue which overlaps the shell above presses the animal firmly against the wood in that region. With the shell held tightly against the head of the burrow in this manner it is easy to see how boring takes place. By the contraction of the hinder muscle, the frontal portions of the shell are drawn apart so that the sharp ridges with which they are covered scrape off the surface of the wood while at the same time the ridges on the middle lobe widen the opening behind. The small muscle between the frontal regions now comes into play and by contracting in its turn brings these together again, which is a much easier operation because the surface of the wood offers no resistance to movement in this direction. The foot then loosens and rotates a short distance in one direction, when the same process is repeated. This goes on time after time until the shell and front half of the body have twisted completely round—the hinder end cannot do this because it is attached to the burrow in the region of the pallets—when the movement is reversed. As a result of this continuous working of the shell and twisting of the front end of the body first in this direction and then in that, the inside of the burrow becomes perfectly smooth and circular in cross section. Immediately behind the shell the surface of the wood becomes covered with a shell layer which is produced by the naked region of the body. The whole process of boring is a model of mechanical efficiency which is not to be surpassed in any other group of the animal kingdom.

The mouth of the shipworm lies just above the sucker-like foot. Into it pass not only the tiny particles collected from the sea water which enter through the inhalant siphon but also the minute fragments of wood which are cut away by the shell. All of these have to pass through the gut of the animal before they can be discharged into the sea water through the exhalant siphon. It was once a matter of dispute whether the shipworm could actually digest the wood or whether this passed through the body unchanged, but there is now no doubt that from this source the animal does obtain at least the energy needed for boring. Wood fragments are acted upon by digestive juices with production of simple sugars. There is, moreover, a large extension of the stomach which is always filled with shavings of wood and seems to be well adapted for the storage of such a slowly-digested substance; it is perhaps also a form of safety valve.

Once encased in its burrow no shipworm can ever leave it. As we have seen, the opening is not much larger than a double pinhead, whereas the burrow widens out within and may, in the common species, be up to 13 mm in diameter. Moreover, the animal is actually attached to the burrow in the region of the pallets. If a shipworm is taken out of its burrow, no matter how carefully this is done, the animal cannot make a new burrow for itself. Since this is the case, how does it happen that new wood becomes infected, as it very quickly does if conditions are favourable for the growth of the shipworm? During spring and summer, eggs and sperms or in some cases larvae are discharged into the sea by way of the exhalant siphon. In this early stage of their existence shipworms are exactly like young mussels or other bivalves. After a short time a pair of tiny shell valves appear which entirely enclose the body and from between which a ciliated sail or 'velum' can be protruded enabling the veliger larvae to swim about, in exactly the same manner as an oyster. It is not known for exactly how

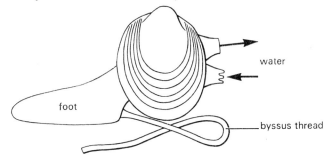

water

foot

byssus thread

Very early settled stage of a shipworm, before boring begins, showing the un-modified shell with on left, the crawling foot and the temporarily attaching byssus thread and on right, the short siphons. (After Sigerfoos)

long the shipworms remain in this state. They probably do so for several weeks and during this time may be carried for great distances by ocean currents or wind drifts. This early stage in their existence is the only time when the shipworms are able to move about freely in the sea and when they are able to infect new timber. It has been shown by experiments that should these larval shipworms chance to come in contact with wood (or indeed with an extract of wood made in alcohol or ether), they remain on its surface whereas they do not remain on any other hard surface such as stone.

Immediately after it has alighted on the surface of the wood, the larval shipworm begins to change; first of all it loses the ciliated velum used for swimming, developing in its stead a long, tongue-shaped foot with which it moves about on the wood until it finds a suitable place to begin boring. This found, it first attaches itself by a fine byssus thread and begins to bore. Both shell and foot are quickly converted into those of an adult shipworm and the animal begins to cut its way quickly into the wood. The pallets and the siphons are formed and remain attached to the burrow near its opening while, as the burrow grows longer and longer, the naked body elongates until the worm-like appearance of the adult is attained. Shipworms usually enter the wood at right angles to the grain but they soon turn in the direction in which this runs and excavate a long burrow. However many animals there may be in a piece of wood, the burrows never run into one another; to avoid this they will twist and turn and interlace with one another in the most intricate manner, as an X-ray photograph of a piece of heavily infected timber shows extremely clearly (Plate 23). Owing to the small size of the openings of the burrows the wood may be heavily infected and show no indication of this fact; finally, however, it crumbles away (Plate 23). Shipworms do not live long, a year or perhaps two years. When they reach a certain age or when there is no further wood to bore into, they continue the shelly casing over the front end of the burrow and remain quiescent within this, taking such food as they can obtain from the water, until they die.

The action of shipworms in northern waters is much slower than that of the more tropical species. In the former, a piece of untreated wood is seldom attacked until it has been in the water for at least a year, whereas in tropical seas wood will become infected in a few weeks and after six weeks show a similar degree of infection to wood which has been exposed in northern areas for eighteen months.

There are many different species of shipworms belonging to several genera. There are three British species, all formerly regarded as belonging to the genus *Teredo* but now designated *Teredo navalis* (the original Linnean species), the larger and commoner *Nototeredo norvagica* and *Psiloteredo megotara* which is confined to floating timber. None is normally over 30 cm long, usually shorter, but there is a tropical 'giant Teredo' (*Kuphus polythalamia*) which forms a massive tube up to 1·8 m long and as thick as a man's arm. Although probably beginning life in wood it comes to inhabit mud amongst mangrove swamps. The largest genus is *Bankia* with species throughout the world, especially in warmer seas. They are distinguished by the possession of long, segmented pallets.

Two other genera of bivalves also inhabit wood. One, *Xylophaga*, the wood piddock, is not uncommon in floating timber in temperate seas. Other species occur in deep water where they may bore into the covering of submarine cables. The shell is very like that of a shipworm and the mode of boring is similar but there is no elongate body; the siphons project from the hind end of the shell which completely encloses the body. The burrows are short, seldom more than 3·8 cm deep, and are not lined with shell. Although the burrows are made as they are in *Teredo* by means of the shell valves, there are no pallets and the animals cannot digest the wood. The same is true of the third type of wood borer, *Martesia*. This is a native of the tropics and is very like a small mussel in appearance; its burrows are not generally more than 6·4 cm long and 2·5 cm wide, i.e. the size of the animal which lives within them. Neither of these animals has attained the efficiency of *Teredo* as a borer and both seek the wood mainly as a means of protection, and not also as a source of food.

There are numbers of Crustacea which habitually bore into wood. Certain of them are pre-eminent, like the shipworms amongst the molluscan borers, on account of their ubiquity and the great damage done. These are the gribbles, species of *Limnoria*, little isopods resembling miniature woodlice, usually between 3 and 4 mm long and with a semi-cylindrical body divided into segments. They have seven pairs of short legs each ending with a sharp, curved claw by means of which the animal holds on to the sides of the burrow. Beneath the hinder end of the body there are five pairs of legs each carrying two broad plates which act as gills; these keep up a continuous movement during the life of the animal and so constantly renew the water needed for respiration. The animal bores into the wood by means of a pair of stout mandibles, one on either side of

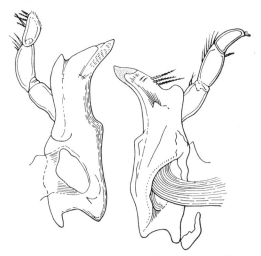

Mandibles of the gribble, *Limnoria*, showing rasp-and-file combination, greatly enlarged. (After Hoek)

the mouth. These are not identical, for the one on the right has a sharp point and a roughened edge which fits into a groove with a rasp-like surface in the left mandible, the whole providing a rasp-and-file combination, as shown here.

Unlike the burrows of the shipworm, those of the gribble are always the same

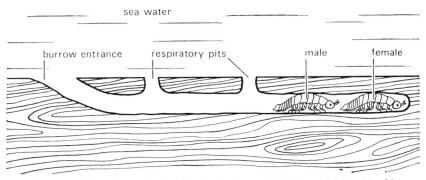

Diagram to show the method of boring of the gribble, *Limnoria*. (Considerably enlarged)

width throughout, so that it appears as though the animals must come out of their burrows as they increase in size and start new ones. Another way in which the gribble differs from the shipworm is that it is always the adult animals which start new burrows. To do this, they first of all hollow out a groove along the surface, keeping in the soft part of the grain, and then pass by a very easy incline into the wood. Usually the burrows are not deep; the need for obtaining a constant supply of fresh sea water probably controls this, and, though they have been found as deep as 15 mm below the surface, they do not usually penetrate for more than a third of this distance. The burrows are often 18 mm or more in length. It is easy to follow the course of a burrow from above because the roof is perforated by a regular series of fine holes like minute 'man-holes' which probably help the animals to maintain the necessary circulation of water within the burrow. The female does most of the work, for, though there are usually a pair of gribbles in each burrow, the female is invariably at the head end. When the males are touched they will crawl backwards slowly out of the burrow, but the females on similar provocation will brace themselves firmly against the side of the cavity by means of their broad tails and successfully resist attempts to pull them out. Probably they brace themselves in this manner when boring, for essentially the same purpose as the shipworm uses its sucker-like foot.

Instead of the myriads of minute eggs which the shipworm produces, the female gribble produces only about twenty or thirty eggs but she takes very good care of these and incubates them in a special brood-pouch beneath her body. There the developing gribbles remain till they have reached a relatively large size and when they finally hatch out they are about one-fifth the size of their parents and are fully formed animals capable of proceeding immediately about their life's business of boring. They do this by hollowing out little burrows from the side of the parent burrow, and *never* by leaving the burrow and beginning a new one in fresh timber. As we shall see later, the fact that fresh timber is always infected by adult and not young gribbles is unfortunate from the point of view of the protection of wood.

The ravages of the gribbles are always clearly apparent on the surface of the wood which is gradually rotted away until the outer layer falls off; the gribble is then able to penetrate still deeper and so, layer on layer, the wood is destroyed. The gribbles are always especially abundant in such structures as pier piles about low water mark and here they eat deepest into the wood, which tapers away and finally breaks at this point (Plate 23). So numerous are they in badly infected wood, that between 300 and 400 gribbles have been collected from a square inch (6·5 cm²) of timber.

There is evidence that gribbles can digest some of the constituents of the wood bitten off by the powerful mandibles. They have been found boring in the insulating covering of submarine cables so that clearly they do not depend on wood to the same extent as shipworms which have never been found anywhere but in wood. Some of the other crustacean borers as well as the mollusc *Xylophaga* have also been found in the insulation of cables and all these creatures appear to bore essentially for protection.

There are several other crustacean wood borers. The most common of these is an amphipod known as *Chelura terebrans* (it has no common name, unfortunately) which is slightly larger than the gribbles, and flattened from side to side, being a relative of the common sand-hoppers of the shore. It usually works along with the gribble, but nearer the surface of the wood, and is almost as world-wide in its distribution. Since it is rarely, if ever, found apart from the gribbles it seems as though it needs the pioneer assistance of these highly competent animals before it can itself attack the wood with any success. In the warmer seas in particular there is a common isopod borer called *Sphæroma* which is rather like a large gribble, to which it is fairly closely related. It often measures 12 mm in length and constructs burrows about 5 mm wide. It has the habit, when disturbed, of rolling itself up into a round ball. It does not do so much damage as the other two crustaceans but can work, as they cannot, in water that is almost, or entirely, fresh.

Rock Borers

In spite of its greater hardness, rock is bored into by a greater variety of animals than is wood and some of these borers can compare with the shipworm in efficiency. Amongst the animals which bore into stone are sponges, worms, molluscs, and crustaceans, while it is also attacked, strange though it may sound, by plants. Limestone rock and especially oyster shells (made of the same calcareous material) are often eaten into by a boring sponge called *Cliona*. The surface of the rock or shell is found covered with many minute holes which lead into branching passages within which the sponge lives. Oyster shells may be completely destroyed in this manner and on some oyster beds boring sponges are a serious pest. They seldom penetrate far into rock, 5 cm at the most. Penetration proceeds by a sequence of chemical and mechanical action.

Several worms spend their lives within tubes which they have hollowed out in rock. Though very small they often occur in great numbers. The tubes are often U-shaped, as in species of *Polydora* which sometimes are common in oyster shells; they may be oval or in the form of a figure of eight. These worms all possess strong bristles which may assist them in boring but possibly the greater part of this work is done chemically because they appear restricted to calcareous rocks.

But, as with the wood borers, the largest and most efficient rock borers are bivalve molluscs. The largest are the familiar piddocks with white spiny shells up to 15 cm long. They are related to the shipworms and bore by the rasping action of the shell valves, the rows of spines gradually cutting into the rock until a boring of anything up to 30 cm in length in the largest species has been excavated. Piddocks bore into a wide variety of substances and are to be found in limestone, sandstone, shale, mica-schist, clay, peat and, very occasionally, in wood. In animals which bore into the softer material the spines are long and pointed; in those taken from harder rock they are round and blunt. During the process of boring, the head of the boring is gripped by the sucker-like foot, as it is in the shipworms, while the shell twists one way and then the other. There are limits, apparently, to the efficiency of this apparatus because piddocks are never found boring in the hardest type of rocks, such as granite. Moreover, although the head end of the boring is certainly cut by the mechanical action of the shell valves, the hind end of the boring, within the opening, also increases in diameter as the animal grows, which would indicate some chemical action because only the fleshy siphons are in contact with the rock in this region. A peculiar feature of the common piddock, the large *Pholas dactylus*, is that although it lives hidden away in rock borings it is nevertheless highly luminescent see Chapter 8).

There is a number of species of piddocks, belonging to the genera *Pholas*, *Barnea* and *Pholadidea*, which are common, while there is a number of rather rarer rock-boring bivalves of which perhaps the most interesting is *Petricola pholadiformis* which, though it belongs to a family of bivalves far removed from the piddocks, has come, as a result of its similar habit of life, to resemble these animals to a striking extent. The commonest of all rock borers is a smaller mollusc, never more than about 2·5 cm long, called *Hiatella* (*Saxicava*) which is abundant both in deep and in shallow waters. Usually it is found in limestone rocks, often in those of such hardness that it seems impossible that so comparatively delicate a shell can have excavated the burrows. It is possible that some amount of the boring is performed by chemical means. No especial glands for the production of acids for eating away the rock have been discovered in either *Hiatella* or the piddocks but it may be that the soft parts of the body which project beyond the shell may exercise a solvent action of some kind.

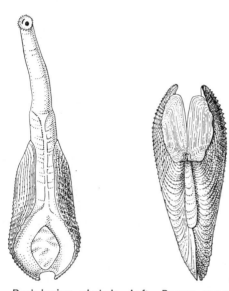

Rock-boring pholads. Left, *Barnea parva* from underside showing extended siphons above and roundish sucker-like foot below. Right, *Pholas dactylus*, common piddock, shell from above showing the secondary plates between the valves, paired 'protoplax' above (anterior) and single 'mesoplax' and 'metaplax' below

There is one rock borer which undoubtedly makes its burrows by the aid of chemicals. This is the date-mussel (*Lithophaga*) of the Mediterranean and tropical seas, so called because of its resemblance in colour, size and shape, to the date. It is a close relative of the common mussel. The brown, date-like colour of this animal is due to the presence of a thick horny layer over the surface of the smooth torpedo-shaped shell. It always bores into limestone (Plate 36), being especially common in dead coral rock, penetrating by the aid of acid produced by glands in tissues which are protruded from between the shell valves and applied to the surface of the rock. When softened in this way the rock can then be scraped by the shell valves, themselves protected from the effect of the acid by the thick, horny covering. Similar glands, although smaller and not producing acid, occur in related, non-boring, mussels. The borings have proved of

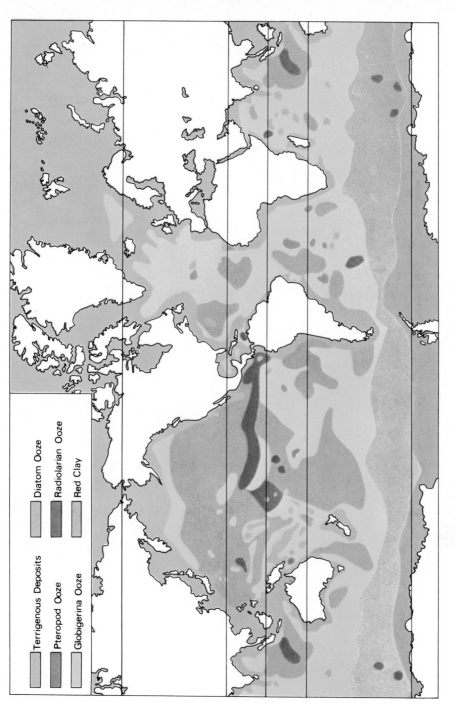

Terrigenous Deposits Diatom Ooze
Pteropod Ooze Radiolarian Ooze
Globigerina Ooze Red Clay

PLATE 17   Map showing distribution of bottom deposits

(above) Blue whale, largest of the whalebone whales, at South Georgia (Photo: N. A. Mackintosh)

PLATE 18

(below) Sperm whale (Photo: Institute of Oceanographic Sciences)

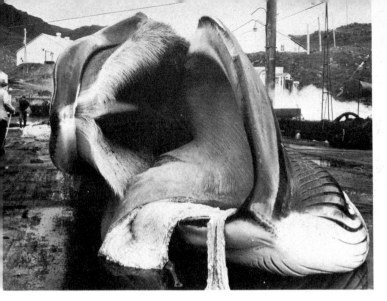

(left) Blue whale
showing baleen
(Photo: Institute of
Oceanographic
Sciences)

PLATE 19

(left) School of
common dolphins,
Delphinus delphinus
(Photo: F. C.
Fraser)

(right) Dolphins,
close-up (Photo:
Institute of
Oceanographic
Sciences)

PLATE 20   DEEP-SEA FISH AND DEEP-SEA ANGLERS

1 *Macrostomias longibarbatus*, $\times\frac{2}{3}$. 2 *Stomias valdivae*, $\times\frac{3}{4}$. 3 *Macropharynx longicaudatus*, $\times\frac{1}{2}$.
4 *Argyropelacus affinis*, $\times 1\frac{1}{2}$. 5 *Ceratias couesi*, $\times\frac{3}{5}$. 6 *Melanocetus johnsoni*, $\times 1$. 7 *Melanocetus kechi*, $\times 1$. (From reports of the *Valdivia* Expedition)

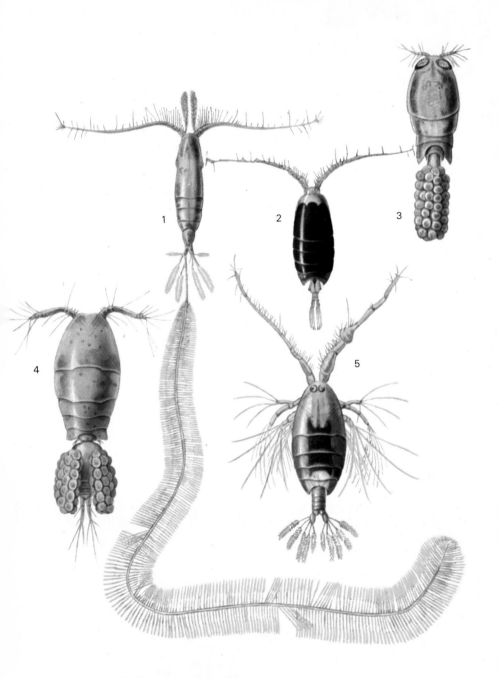

PLATE 21   MEDITERRANEAN COPEPODS

1 *Calocalanus plumulosus*, × 26. 2 *Candace ethiopica*, × 10. 3 *Corycaeus venustus*, × 39. 4 *Oncaea mediterranea*, × 39. 5 *Pontellina plumata*, × 20 (From *Fauna und Flora des Golfes von Neapel* by kind permission of the Director, Stazione Zoologica, Naples)

(right) A crustacean copepod, *Calanus finmarchicus*, × 12 (Photo: D. P. Wilson)

(below) Krill, a crustacean euphausiid, *Meganyctiphanes norvegica*, × ca. 2 (Photo: J. D. Mauchline)

PLATE 22

(below left) Plankton with dinoflagellates, *Ceratium tripos*, × ca. 45 (Photo: D. P. Wilson)

(below) Spring phytoplankton, × ca. 40 (Photo: D. P. Wilson)

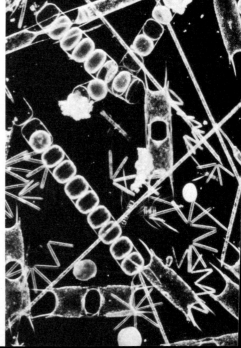

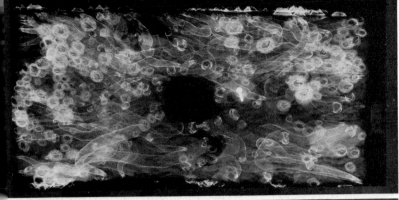

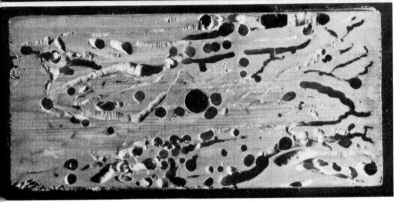

PLATE 23

(above) Timber infected by shipworm *Lyrodus (Teredo) pedicellatus*. (upper) X-ray photograph showing animals *in situ*. (lower) Cut surface showing borings, ×$\frac{2}{3}$ (Photos: Central Electricity Research Laboratory)

(right) Piles of an old jetty at Penrhyn Bay, N. Wales eaten away by *Limnoria*, the gribble. Shortly afterwards the jetty was washed away in a storm (Photo: D. P. Wilson)

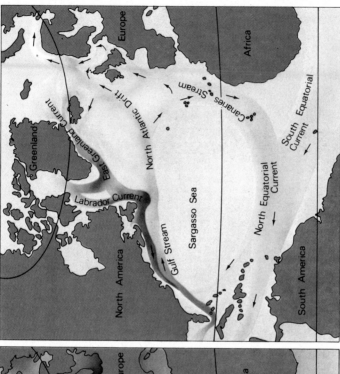

Chart showing distribution of plankton in North
Atlantic
Darkest green, most abundant. Blue, scarcest

Surface currents in North Atlantic
Red, warm currents. Blue, cold currents

PLATE 24

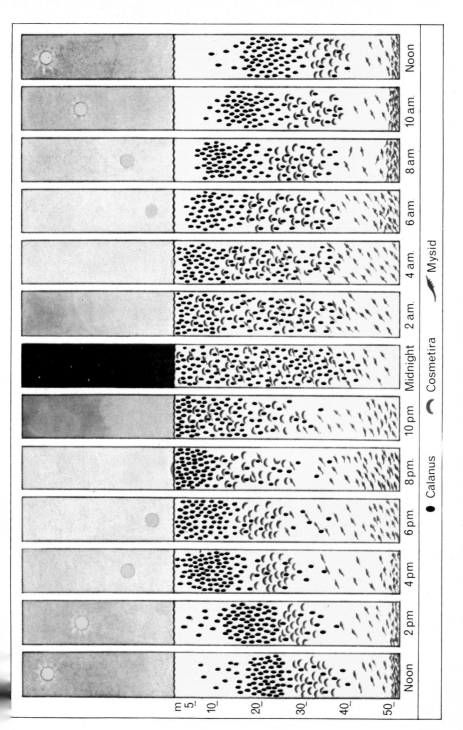

PLATE 25   VERTICAL MIGRATION OF PLANKTON ANIMALS

Showing behaviour of a crustacean copepod, *Calanus finmarchicus*, a medusa, *Cosmetira pilosella*, and a crustacean mysid, *Leptomysis gracilis*, which lives on the bottom in the daytime (After F. S. Russell)

● Calanus   (Cosmetira   ✦ Mysid

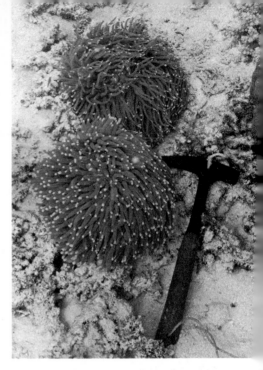

(*above*) Giant clam, *Tridacna gigas*, about 1 metre long living at a depth of some 8 metres on Kossol Reef, Palau Isles, West Pacific (Photo: T. F. Goreau)

(*above*) Large mushroom or fungid corals, *Heliofung. actiniformis*, with long tentacles characteristic of th species (unlike other fungids). Same locality, about i metres depth (Photo: T. F. Goreau)

PLATE 26

(*below*) Crown-of-thorns starfish, *Acanthaster planci*, on stags-horn coral, species of *Acropora*; white area on left recently destroyed. Low Isles Reef, Great Barrier Reef of Australia (Photo: T. F. Goreau)

PLATE 27

(above) Exposed outer crest of coral reef cemented by calcareous litho-thamnid red algae. North-west Island Reef, Great Barrier Reef (Photo: M. J. Yonge)

(centre) Exposed coral, largely stagshorn, on lee of Pixie Reef, Great Barrier Reef (Photo: M. J. Yonge)

Dark green patches composed of dense populations of the flatworm, *Convoluta roscoffensis*, in damp patches on sand below rock covered with fucoid weeds. Roscoff, Brittany (Photo: C. M. Yonge)

Indo-Pacific reef-
building corals.
Branching *Stylophora*
in foreground, rounded
Goniopora behind
(Photo: T. F. Goreau)

PLATE 28

Atlantic reef-building
corals. Encrusting and
plate-like species of
Agaricia (Photo: T. F.
Goreau)

PLATE 29 Red coral of commerce, *Corallium rubrum*, showing expanded polyps with eight pinnate tentacles. 1 and 2 Pieces $2\frac{1}{2}$ months old, about natural size. 3 Portions of colony with polyps expanded, natural size. 4 Polyps retracted, slightly enlarged. 5 Portion greatly enlarged with polyps fully expanded. (From *Histoire naturelle du Corail* by H. de Lacaze-Duthiers. Paris, 1864)

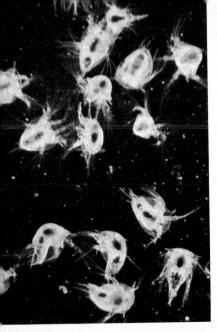

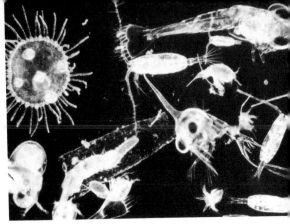

(above) Plankton containing larval crustacea and worms, and a medusa, Obelia, × ca. 6 (Photo: D. P. Wilson)

(left) Nauplius larvae of acorn barnacle, *Balanus*, × ca. 18 (Photo: D. P. Wilson)

PLATE 30

Swell waves (Pho Underwood Press)

Waves breaking on sh (Photo: Underwood Pre

 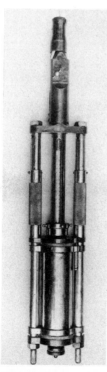 

(above left) Drift bottles. *(left)* Surface bottle, × ca. ⅓ *(right)* Bottom trailer with wire to keep bottle off bottom, × ca. ⅑ *(above right)* Plastic water bottle in use (Photo: Institute of Oceanographic Sciences)

(above) Insulated water bottle, × ca. ⅛. *(left)* Open. *(right)* Closed

PLATE 31

(above) Plastic drifters on sea bottom (Photo: Ministry of Agriculture, Fisheries and Food)

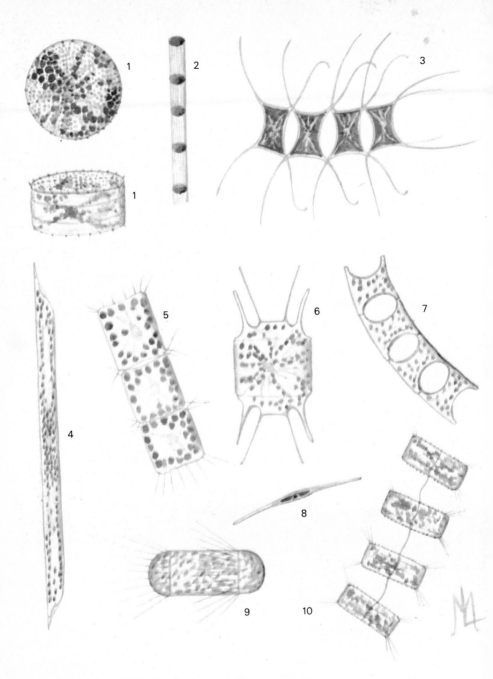

PLATE 32 DIATOMS, all greatly enlarged

1 *Thalassiosira eccentrica*. 2 *Skeletonema costatum*. 3 *Chaetoceros curvisetum*. 4 *Rhizosolenia styliformis*. 5 *Lauderia borealis*. 6 *Biddulphia mobiliensis*. 7 *Eucampia zoodiacus*. 8 *Nitzschia closterium*. 9 *Corethron criophilum*. 10 *Thalassiosira gravida* (After M. V. Lebour)

great value in a somewhat unexpected connection, that of earth movements. The classic example is the borings in the limestone pillars of the 'Temple of Seraphis' at Pozzuoli near Naples. These pillars are today some distance above the sea, but around a certain level are perforated with the burrows of the date-mussel. Since the temple must originally have been built above sea level, it is clear that the land sank until the pillars were covered at least to the height to which they have been bored into. It is just as obvious that since that time the land has risen again because the temple is once more on dry land. Thus the presence of the holes made by the date-mussel enables us to prove that the land in the neighbourhood of Pozzuoli has both sunk and risen again within historic time.

Crustaceans do far less damage to stone than to wood. Coral rock is bored into by species of the barnacle, *Lithotrya*, but the only other crustacean which does much damage is another species of the *Sphæroma* which bores in wood. This animal can only penetrate soft sandstone or claystone but has done considerable damage in New Zealand by destroying harbour works made of the latter.

The rock-boring animals cannot obtain food from the rock; the advantage of boring is purely protection. The borings are seldom long and the animals are always in free communication with the water outside, from which they take food, consisting of minute planktonic plants. In general, therefore, they resemble all the molluscan wood borers with the exception of the remarkably specialized shipworm which is unique amongst borers in its inability to make a new burrow if removed from the one it originally constructed, and its capacity to obtain energy from the substance into which it bores.

Besides being attacked by animals, rocks, surprisingly enough, are also bored into by marine algæ. Here again it is only limestone or other calcareous rocks which are attacked, and the plants appear to prefer the harder varieties which, by their action, are soon reduced to a friable mass, coloured blue or greenish owing to the plants everywhere present in their substance. Their preference for calcareous rocks is due to the fact that, like the date-mussel, they bore by chemical means, transforming the hard carbonate of lime which composes the rock into the soft, powdery, white bicarbonate. Besides rocks they attack the shells of molluscs, doing great damage to the oyster beds in the Bay of Sebastopol in the Black Sea where they are especially common. In the same regions they attack the shells of barnacles, the calcareous tubes of worms and the external limy skeletons of other marine creatures, all of which are coloured by them. Their prevalence in the Black Sea, where they are common in all depths down to about 23 m, is a serious economic problem on account of the damage they do both to stone work and animal life. Fine strands of green algæ penetrate the substance of coral rock but without doing appreciable damage.

Protection of Timber

The problem of the protection of timber from the attacks of marine borers has vexed mankind since the days of the earliest wooden ships and harbour works. Today, though the advent of steel ships and concrete harbour works has led to a great reduction in the use of timber in marine structures, the problem remains

H

unsolved. How urgent is the need for a solution can be judged by the extent of the damage done in San Francisco Bay following a great burst of activity by ship-worms in that district between 1914 and 1920, when wooden piers and wharves to the estimated value of ten million dollars were destroyed.

Innumerable remedies and protective measures have been tried with varying degrees of, but never complete, success. Different woods vary with regard to the speed with which they are attacked. Hard woods, such as teak or oak, are attacked just as quickly as softer woods, but some woods resist attack for a longer period owing to the presence in them of essential oils or poisonous alkaloids. Greenheart and eucalyptus are examples. One of the methods of protecting timber consists of leaving the bark intact, for this often contains such protective oils. The wood, however, requires very careful handling for if the bark is injured the wood borers make their way through the damaged places. In any case there is always the danger that the animals will enter about the knots in the wood.

It is impossible here to do more than give a list of a few of the more important methods of protecting timber. Protective measures fall naturally into two main divisions, those which protect the surface of the wood and those which impreg-nate it with poison. Piles have been painted with a great variety of substances in the hope of keeping out borers; amongst them may be mentioned tar, copper paints, and innumerable patent preparations. These are usually efficacious while the coating remains intact, which, in the case of pier piles that are continually being scraped by the sides of ships, is not long. Charring or 'breaming' of timber is another old method of protection, but is not, apparently, very effective. Then there is a large number of methods which aim at the protection of timber by armour of one kind or another. Metal has been largely used for this. Steel or iron sheathing has been employed but found to corrode too quickly, zinc and Muntz metal (an alloy of copper and zinc) also giving indifferent results; much better are casings of cast iron or copper. The latter is highly poisonous to all forms of life but is expensive and, unfortunately, much sought after by thieves. Vitrified pipe castings have been found to give excellent protection but they, in their turn, suffer from the serious defect that they are easily broken. A great number of patent armours of one kind and another has been put on the market but, though they give initial protection, all break down sooner or later and the wood borers enter. Concrete casings are largely used at the present day, especially in the United States, and they are often very effective but there is always the danger that the concrete will be attacked chemically by the sea water or that it will break away near the base of the pile and so allow borers to effect an entrance there and work their way upwards. An old and very effective method consists of cover-ing the surface of the wood with broad-headed iron nails, known as 'scupper-nailing'. The rust from the nails spreads and forms a complete coating over the surface with satisfactory effect. All these methods are more effective with the crustacean borers than with shipworms, since the former can do little damage unless a good deal of the surface is exposed, whereas a very small unprotected area will allow comparatively large numbers of minute larval shipworms to begin boring, and these may each penetrate the wood for a depth of 30 cm or more, according to the particular species.

The best method of protecting timber against the ravages of shipworms is to impregnate it with some poisonous substance. Scientific investigations into the most suitable means of affording this type of protection have been conducted both in Britain and in the United States over many years, initially by the Sea Action Committee of the Institution of Civil Engineers in the former and by a Committee on Marine Piling Investigations in the latter. After exhaustive tests of a great variety of substances, various organic compounds containing arsenic, some of which were used as poison gases during the first world war, have been found the most efficacious, and experiments have been carried out with these substances, by exposing in the sea rafts containing timber treated in various ways, and finding the degrees of protection afforded by the various poisons. In all cases it is necessary to introduce them into the wood in solution in creosote, otherwise they wash out and the wood is left fully exposed to infection. The creosote alone, if the timber is well impregnated with it, often provides good protection for a long period. The wood has to be thoroughly dried before impregnation, the latter process being carried out either by the aid of a vacuum which draws the creosote through or else by boiling with the impregnating fluid in an open tank. Up to date these experiments are having good results so far as shipworms are concerned but are not so efficacious with the gribbles, which appear to be less susceptible to poisons. This is probably to be explained by the fact that only fully grown gribbles attack new timber and these will not be so easily poisoned as the minute larval creatures which are the carriers of infection in the case of the shipworm. Other protection measures include the effect of detonation which kills the animals in their borings.

The rock borers do not do a fraction of the damage of the wood borers and are a minor problem. Mechanical borers, such as piddocks, may do a good deal of damage to soft rocks, and calcareous rocks are always in danger from the chemical borers, like the date-mussel, but if these types of rock are avoided there is no danger. Moreover they work much more slowly and penetrate far less deeply than do the wood borers. Cement and concrete are seldom attacked by them, though there are a few recorded cases of these substances being bored into by piddocks or their relatives.

Chapter 7
CORAL REEFS

The last chapter was concerned with animals and plants which are agents of marine destruction; here we are concerned with those which serve the opposite function being responsible for the formation of innumerable coral reefs and islands in tropical seas. With the advent of Scuba diving and of underwater colour photography, all are now aware of the complexity and beauty of life on these marine formations. Here marine biology joins hands with geology and geography because animals and plants are directly responsible for major additions to, and modifications of, the land surface. Although a wide diversity of organisms is responsible for the formation of reefs, the two most important are animals closely allied to sea anemones and red seaweeds related to the little corallines common in temperate rock pools. By the slow but never ceasing conversion of calcium salts in solution into calcareous rock, dry land may eventually appear where formerly was open sea, land on which first plants, then animals and finally possibly man himself, have successfully established themselves.

Coral reefs occur only in shallow tropical seas, descending to no more than some 90 m with the temperature not dropping below about 18·5°C. They extend beyond the tropics only into regions of unusually high temperature such as the northern ends of the Persian Gulf and Red Sea. The Gulf Stream is responsible for the northern extension of some coral growth to Bermuda. The populations of reefs in the Indo-Pacific and the Atlantic are distinct with a greater diversity, both of corals and other organisms, in the former oceanic region. Reefs have been described by T. F. Goreau, the most distinguished of recent workers on them, as consisting (Plate 27) of a primary framework of a 'cemented mass of large interlocking coral and algal colonies buried in their original position by subsequent overgrowth on all sides'. At their bases, in deeper, less illuminated water, secondary or detrital reefs extend. With the primary reef-building corals and calcareous red seaweeds, are secondary 'framefillers' which would also include some large molluscs, and a host of smaller animals and plants all with calcareous skeletons which break down to form the sand which everywhere fills in cavities and extends over the bottom, especially in the lee of reefs. All this adds up to a highly complex structure, but if we think of the larger coral skeletons as providing the bricks and the encrusting sheets of pink calcareous algae as forming the essential binding cement, we shall have some basic idea of how a reef is formed.

We can now turn to consideration of corals as animals. The term coral has been widely applied to all animals—and plants—that form massive calcareous skeletons. Here we shall confine it to the stony corals, members of the order Scleractinia (formerly Madreporaria) with just a few related coelenterates. The stony corals are very closely related to the sea anemones (Actiniaria) but with the added capacity of forming a massive limy skeleton. Moreover, they are usually colonial, some species forming massive rounded and branched colonies many metres across and weighing many tonnes. The simplest corals, however, are solitary individuals like the small cup corals, *Balanophyllia* and *Caryophyllia*, of temperate seas. Looking down on these when expanded, with the underlying skeletal cup obscured by the extended tentacles, they appear to be anemones. But on withdrawal they contract into the cavity, or calyx, of the skeleton. This consists of a basal plate, and a surrounding wall from which a series of partitions, the septa, extend inwards leaving a small central cavity. When studying the skeleton it must be remembered that the animal or polyp rises above it when expanded for feeding and digestion. This process occurs largely at night when the majority of corals expand, remaining contracted and pulled down hard against the underlying skeleton during the day.

The presence of these septa distinguishes the scleractinian corals from the hydroid and the alcyonarian corals some of which are also important reef-builders. The former are allied to the small hydroids so common in shallow temperate seas. One of them, *Millepora*—known as the stinging coral owing to the power of its sting cells, the threads from which can penetrate human skin—occurs on coral reefs all over the world. The smooth surface of the skeleton is perforated with series of fine apertures, each consisting of a central opening surrounded by a ring of smaller ones. From the former extends the feeding polyp with fine tentacles and central mouth, from the others protective tentacles armed with sting cells but without mouths. The alcyonarian corals are allied to the dead-men's-fingers (*Alcyonium*) of British coasts and the precious coral (Plate 29) of warmer, but non-tropical, seas. The tentacles are branched and are eight in number (those of anemones and scleractinian corals are simple and in multiples of six). There are two important alcyonarian corals, both confined to Indo-Pacific reefs. There is the well-known red 'organ-pipe' coral, *Tubipora musica*. The deep red of the skeleton is obscured in life by the emerald green tentacles projecting from the free ends of the parallel upright tubes of which the skeleton consists and which do resemble organ pipes. Only the surface region of the coral contains living tissues, the underlying skeleton, with the parallel 'pipes' united by horizontal partitions, becoming a purely supporting structure and, incidentally, the home of innumerable worms, crabs, sponges, sea mats and many other animals and plants. The skeleton is not solid, as it is in the scleractinians and in *Millepora*, but is composed of tiny fused spicules such as are found embedded in the gelatinous substance of typical alcyonarians, the so-called 'soft corals' which are widely abundant on Indo-Pacific coral reefs. The other alcyonarian coral is an important reef-builder which *Tubipora* is not. This is the blue coral, *Heliopora*, which forms massive colonies often some metres in diameter, in shallow water. The polyps here are minute and extremely difficult to observe. Only pale blue on the surface, when broken the interior of the skeleton is deeply coloured. Again it is composed of many, in this case very fine, parallel tubes but there are no spicules; the skeleton is a solid mass of crystalline calcium carbonate.

There is a further, highly significant distinction to be made concerning corals. The stony scleractinians occur in all seas and at all depths down to the abyssal. It is only in the shallow tropical waters that they form massive reefs capable of with-standing the full force of the sea. We therefore divide them into non-reef-building (or ahermatypic) and reef-building (or hermatypic) corals. This is a distinction based on habit *not* on structure. Some families have both ahermatypic and hermatypic members. The only structural distinction is the greater number of polyps in a given skeletal area in the reef-builders. Other distinctions will emerge later.

The calcareous seaweeds that form the essential binding cement are mainly red algae, appearing pinkish in life but bleaching dead white after death. The commonest are the lithothamnioid algae (Plate 4) species of which occur all over the world. An important green weed which deposits carbonate of lime in its fronds is *Halimeda* the numerous species of which form an important constituent of many reefs. It has a characteristic appearance being formed of a number of broad limy segments united by narrow, uncalcified joints which render the plant relatively flexible. On death the lobes separate and form a significant proportion of coral sand.

The fact that corals are responsible for forming often enormous masses of calcium carbonate must not blind us to the fact that they are animals. We have to consider such matters as their manner of reproduction and growth, their adaptations for feeding and for disposing of the sediment which continually falls on to them and the varied means whereby different species are suited for life in all the varied environments—of shelter and exposure—presented on a reef. Coral colonies grow by a process of repeated division of polyps, and so of the calyces in the skeleton in which they lie and which they also form. The process, very similar to what happens in the growth of a plant, is known as vegetative propagation. When the corals are sexually mature, eggs are liberated into the gastric cavity (coelenteron) where they are fertilized by sperm from other polyps or colonies which is liberated into the sea and so carried into the female polyps. There the fertilized egg develops to form a round or oval larva known as a planula (exactly as in sea anemones). These planulae are covered with cilia and at this stage—each about the size of a pin's head—they are discharged, a process known as planulation.

A period of some days is spent as a temporary member of the plankton before the larva reacts to gravity instead of light and so drops to the bottom. There it settles should it find a suitable clean, hard surface. It now suffers the metamorphosis that occurs in all larvae when they change into the adult form. The under-surface spreads out to form a broad basal disc while the upper surface is pushed in to form a mouth and gastric cavity and around the mouth a ring of tentacles appears. The skeleton now begins to form, first a basal plate, then a surrounding wall with the first formed septa growing inward from this. If a solitary coral, only further increase in size follows, but if it is a colonial, reef-building coral it is only in the very earliest stages of its development. Division begins after the fashion shown below, small individuals budding off from the side of the first-formed polyp. This may then continue to grow upward budding off side polyps as it goes, the original polyp forming the apex of the whole colony. In another type of coral growth, the apical polyp, instead of being the original and oldest one, is the youngest, having been budded off from the polyp immediately beneath it and shortly to be superseded when it produces a bud in its turn. Many other corals have no special point of growth, all the polyps having equal powers of division and budding, and in these cases the colony spreads in a regular fashion over the surface of stones and underlying dead coral, finally forming a hemispherical mass such as that usually assumed by the common reef-builders *Porites* and *Favia*. In some cases the division of the polyps may not be complete; instead of the surface of the colony having many openings, all distinct from one another, it may be covered with wandering, sinuous grooves fringed with the

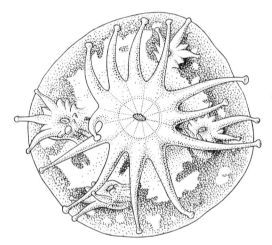

Recently settled coral, *Pocillo-pora*, showing first formed individual with central mouth and surrounding tentacles and new polyps appearing around it; calcareous basal plate spreading over surface below. (After Stephenson)

characteristic septa and representing the site of polyps which have extended themselves, as it were forming buds but never parting company with them. This type of coral is best exemplified by the well-known Meandrines or brain-corals, so called because the meandering depressions with which the surface is covered resemble very closely the convolutions on the surface of the human brain.

There are remarkable solitary corals, common on Indo-Pacific reefs, which are worthy of a short description, so peculiar is their manner of development. The swimming embryo settles and develops into a little cup-shaped coral, but, on attaining a certain size, this swells out at the top until it looks almost exactly like a mushroom turned upside down so as to expose the gills (hence the common name of these fungid corals). After a time this flattened upper portion falls off and is caught in water movements which deposit it in some sheltered pocket where sand collects and where it continues to grow until it may be 15 cm across (Plate 26). The whitened skeletons of these stalkless discs, consisting of a circular depression with radiating septal partitions, are common objects on coral shores and in

zoological museums. The original polyp does not die when the flattened head
falls away but grows a new one which, falling off in its turn, is succeeded by an
indefinite number of others.

There is a great diversity of genera and species of reef-building corals. Many
are easy to distinguish from each other and may be characteristic of particular
regions, from shelter to full exposure, on the reefs. They are adapted for that
particular niche in the environment. The stagshorn coral, *Acropora*, which is the
most widespread of all coral genera with a great number of species in both the
Indo-Pacific and the Atlantic, has species with very delicate branches which
occur only in sheltered areas, others with stout, round-ended branches which live
in exposed conditions and others which grow as encrustations over areas where
the surf breaks. But in some cases, with *Acropora* and other genera, it becomes
difficult to know whether we are dealing with a species or a growth form. Here
again corals resemble plants which may, for instance, have a luxuriant growth in
the valley but become stunted with a very different appearance when grown on
the mountainside. In the case of corals, the chief factors influencing growth form
are the depth at which they live, with consequent reduction in light, the degree
of motion in the water and the amount of sediment. Experiments carried out
during the course of the Great Barrier Reef expedition in 1928–9, when coral
colonies were divided and the two halves then placed under different environmental conditions, showed great difference in subsequent growth and appearance. It will take a great deal of such experimentation to determine whether
apparent species are true species or are growth forms, moulded by the effects of
the environment.

The effect of sediment is of the greatest importance. Coral polyps expand and
contract, and all exposed surfaces are covered with fine, rapidly moving cilia, the
action of which helps to cleanse the surface, but these are their only means of
protecting themselves from sediment. Like all attached animals, they are essentially at the mercy of material which drops on them from above or collects around
them. A branching coral has less to fear from falling sediment than a massive
coral which offers a wide surface on which this can collect. The effect of this is
finally to kill, by a kind of suffocation, the polyps in the centre; growth consequently ceases in this region but is continued round the edges where the polyps
are uninjured, the sediment having slipped off, and so the colony changes from a
globular mass into a great, shallow basin. Another way in which sediment may
affect coral colonies is by accumulating around their bases and killing the lower-
most polyps.

Corals, like sea anemones, are specialized carnivores seizing animal prey,
usually members of the animal plankton, by means of tentacles armed with
batteries of stinging cells. This plankton is most abundant in surface waters by
night when the bulk of the corals are fully expanded. The paralysed prey is
passed into the gastric cavity where it is wrapped around by filaments which
rapidly digest it. Plant material is not accepted and, if placed in the gastric cavity,
is not digested. In some corals where the tentacles are short, food may be carried
into the mouth by cilia which normally beat outwards carrying sediment away.
But on the stimulus of animal material, the cilia in that region change their beat

until the food is swallowed when they resume their normal outward beat. This is true of certain mushroom or fungid corals.

This description of food and feeding brings up the question of the minute algal cells which are invariably present in the tissues of reef-building—but *never* of non-reef-building—corals. These are spherical brown bodies, each about one-hundredth of a millimetre in diameter and are contained within animal cells but only in the cell layers that bound the gastric cavity—the endoderm in contrast to the external cell layer or ectoderm. The nature of these algae was long a mystery and, in view of their colour, they were called zooxanthellae. It is now known that they represent vegetative stages in the life-history of the dino-flagellates (Plate 9) which form so significant a section of the plant plank-ton. When cultured outside the body of corals in suitable media motile stages appear, each with the two flagella, one terminal and one in a groove round the middle, characteristic of the dinoflagellates. Their numbers are astrono-mical; even a planula no larger than a pin's head may contain over seven thou-sand!

The significance of the association is a matter of the first importance. The plants are really 'imprisoned plant plankton' having the advantage of protection within the animal tissues and immediate access to the carbon dioxide needed for formation of carbohydrates by means of their contained chlorophyll, and of nutrients containing the nitrogen, phosphorus and sulphur which are needed for the further elaboration of proteins. These substances are all produced by the animal in the normal chemical activities, or metabolism, of the body. In other animals, including the non-reef-building corals, these substances are excreted into the surrounding water where they are utilized by the plant plankton. This imprisoned plant plankton takes these supplies at source.

From the standpoint of the animal this is a great advantage; the plants repre-sent an automatic means of ridding the body of waste matter—all the more important here because corals, like all coelenterates, have no organs of excretion; waste matter must just diffuse out as it does in the non-reef-builders. This general increase in efficiency may be one reason why reef-builders form the skeleton at a far higher speed. There is no doubt that this happens; skeleton formation (calcification) takes place at a much higher rate by day (when the plants are photosynthesizing in the presence of light) than by night. It is possible to remove the plants by keeping suitable corals in darkness for long periods, after which the rate of calcification falls, being no greater by day than by night. This matter of increased skeleton formation is fundamental. It represents the basic difference between reef-builders which form reefs capable of resisting the full force of oceanic seas, and the non-reef-builders which at best (as in the depths of Norwegian fjords) can only form branched interlocking masses of no structural strength. One sees now, moreover, why it is that reef-builders occur only in shallow, illuminated water whereas other corals occur at all depths. But why reef-builders should be confined to the tropics is still unknown.

There remains the question of whether the algae supply, as many have thought, food for the coral. This has been debated for very many years without any clear answer yet emerging. Corals are, as we have seen, very well equipped for dealing with animal food and will not react to plant material. However, it is now known that organic matter does pass from the interior of algal cells to the exterior, from the interior of free living dinoflagellates into the sea and from the bodies of imprisoned zooxanthellae into the tissues of corals. What still remains to be determined is the precise significance of this organic matter in the living processes of the animal.

It is difficult to estimate the growth of corals, for this depends so much on conditions, especially temperature and the supply of food, but it has been found that the branches of *Acropora* may increase by 2 to 5 cm in length per annum, while a mass of *Porites* increased its diameter by 76 cm in twenty-three years. Other estimations have shown that a shallow-water reef may, under favourable conditions, grow upwards at a rate of about 30 cm in eleven and a half years, or 29 m in one thousand years. In Samoa, Dr. Mayor came to the conclusion that the corals added some 840,000 pounds (381,020 kg) of limestone to the reef annually, although about four times this amount was removed in the same period by currents and by sea cucumbers and by boring animals. This is not the only evidence we possess which indicates that coral reefs, in some areas at any rate, may be decreasing rather than increasing.

The red coralline seaweeds (Plate 4) grow as sheets, usually comparatively thin, although they may form masses some metres thick. These encrusting sheets of pink-coloured cement—they look like nothing else—occur in greatest abundance on the most exposed reef crests (Plate 27) on which the full force of oceanic surf breaks providing the exceptional degree of oxygenation these calcified plants clearly need. Such conditions are provided on the seaward summits of barrier reefs which face the open ocean and on the faces of atolls exposed to the north-east or south-east Trade Winds. The growth of the coralline algae is profoundly influenced by the conditions under which it lives. It forms, under extreme exposure, a series of buttresses or spurs with deep intervening grooves or surge channels (see below) which extend to depths of up to 20 m. The effect of this mode of growth is to dissipate the enormous force of the oceanic breakers which rush up the parallel series of grooves. The landward end of these is frequently overarched by algal growth from the two sides leaving occasional openings through which the water finally shoots high into the air.

Apart from the primary reef formation by corals and coralline algae, many of the larger mollusc shells may become important constituents of the reef mass. This is notably true of the Indo-Pacific giant clams (species of *Tridacna*) (Plate 26) which grow to recorded lengths of over 1·2 m weighing many kilograms and representing the largest bivalve molluscs ever evolved. They also contain symbiotic zooanthellae.

There is constant destruction as well as growth. Corals are very susceptible to accumulation of silt while a variety of animals prey on them, notably the parrot fishes (Scaridae), which scrape the living surface with massive teeth shaped like a parrot's beak, and also the file fishes. Some polychaete worms eat coral polyps but the major predator is the crown-of-thorns starfish (*Acanthaster planci*) which recently has increased to a spectacular extent in many regions of the Indo-Pacific (Plate 26). This crawls slowly over the living surface, everting the stomach and digesting all living tissue as it goes. Whole reefs may be effectively destroyed in this way. The cause of this population explosion remains uncertain although it has been ascribed to a variety of man-made causes such as destruction of reefs providing settlement surface for the young starfish, collection of large snails which feed on the starfish or the effect of various pollutants. However, it could be a cyclical phenomenon such as we are increasingly finding with the growth of knowledge about the long term course of events in the sea.

Dead coral rock is subject to agents of both biological and mechanical erosion, the former often assisting the latter. A variety of bivalve molluscs, such as datemussels and piddocks, bore into the rock, as do a variety of worms, some sponges and even a barnacle (*Lithotrya*). Filamentous green algae also ramify through it. Further breakdown into sand and then mud is assisted by the large sea cucumbers which continuously swallow coral fragments, removing all digestible matter and gradually reducing coarse to ever finer particles.

A reef of living coral (Plate 27) is literally a sea garden. The corals themselves

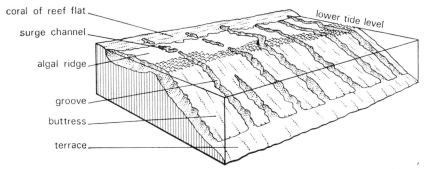

Sketch of seaward face and summit of reef on the windward side of an atoll. Explanation in text. (After Munk and Sargent)

are of all colours, many being brown with violet, pink or white polyps, others orange, green or yellow; the polyps of some corals are foliaceous with long tentacles while others are velvety masses with tentacles almost absent (Plate 28). And the other animals have colours just as vivid. There are many fish of the most varied and brilliant hues which dart hither and thither in little shoals amongst the trees of the coral forest; there are the many-coloured sea cucumbers, large tube-worms with brilliantly coloured crowns of tentacles, large sea anemones, innumerable crabs and all manner of other crustaceans, many of which live in definite association with the corals, a like variety of molluscs and all manner of encrusting organisms such as sea mats and sponges.

As in temperate seas there is a zonation of intertidal and shallow sublittoral life on coral reefs although species are so numerous that interpretation is often very difficult. To survive life on the summit of reefs, particularly when near a land mass, corals and other reef animals must have the necessary functional ability to withstand extremes of temperature and salinity together with the ability to withstand the effects of desiccation. The precise areas occupied by particular corals (excluding their adaptations for life in exposed or more sheltered conditions) depends on a variety of factors. One is certainly the conditions under which particular coral larvae, the planulae, will settle. Another—surprisingly enough—is direct action by one coral against another. Recent work by Dr. Judy Lang in Jamaica has revealed the presence of a 'pecking order' among certain related reef-building corals. When one higher in the order comes, by growth, into physical contact with one lower down, it extends the mesenterial filaments out of the mouth and proceeds literally to digest the tissues of the other coral. Such behaviour must have a highly significant effect on the final distribution of corals on a reef.

Three types of reef were early distinguished. *Fringing Reefs* edge the shores of islands and continents with no deep-water channel between them and the land. *Barrier Reefs* also occur off land masses but at much greater distance and separated from them by a deep and wide channel. The supreme example is the Great Barrier Reef of Australia which extends for over 1930 km, the outermost reefs in places over 160 km offshore. *Atolls* are not connected with any land and consist of rings of coral reef with occasional islands of purely coral formation. Within is a lagoon, which, like the channel within a barrier reef, seldom exceeds 60 m in depth and varies in size between that of a small lake and of an inland sea up to 64 km across. There is usually a moderately deep opening on the lee, that is on the side opposite to the one exposed to the Trade Winds. We now also recognize a fourth category of *Patch Reefs* such as those that grow in great numbers within the shelter of atoll lagoons or within barrier reefs.

The formation of fringing and barrier reefs is clearly bound up with the structure of the land they border, to the constitution of which, however, they do not contribute, although, as the result of local raising of the land, dry land in many regions is composed of coral limestone, originally formed in the sea and now in some cases upon mountains some thousands of metres above the level of the sea. The formation of the real 'coral islands' or atolls is not associated with land, from which they may be separated by hundreds of kilometres, any land present being purely of coral origin.

An explanation of the formation of coral reefs is best begun by examining the manner in which fringing reefs are formed. There is here no such mystery as has surrounded the formation of barrier reefs and still more of atolls. Where there is a suitable bottom, usually rocky, corals will establish themselves around the coasts of continents and rocky oceanic islands. They then proceed to grow upward in the manner shown here. When they reach low-water mark, upward growth ceases because corals cannot withstand exposure to the air for any but very short periods. Unable to continue upwards, the corals must grow outwards, and so the coral mass becomes broader and also much steeper on its outer slope. Growth is reduced in the lowered illumination at the base of the reef, while the mass above continues to grow till a steep precipice is formed on the outer side of the reef down which fall coral boulders and debris of all kinds, forming a detrital reef which in the course of time forms the foundation for further outward extension. The action of the sea, whose breakers unceasingly beat against the edge of the reef, causes portions of coral, shells and the like, to be thrown on to the reef flat, as the flattened upper surface of the reef is called. As a result, the outer edge is often marked by the presence of a long mound which projects above the surface of the sea.

The region immediately below water level on the outer edge of the reef is the area of most active coral growth. On the other hand, the reef flat between the outer zone and the land becomes hollowed out to form a shallow, lagoon-like channel, sometimes deep enough for the passage of native boats. It is not known definitely exactly how these channels are produced, but probably the principal agencies at work are the scouring action of the sea as the waves rush over the surface of the reef flat, the action of the multitudinous boring bivalves, worms, sponges and so forth and, much more doubtfully, the direct dissolution of the

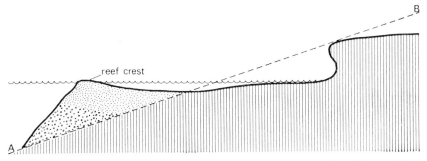

Diagram showing mode of formation of a fringing reef. A–B, original slope of shore, reef (stippled) growing seaward over substrate (vertical lines), shelf so formed extended by erosion. A shallow lagoon lies within the outer boundary of the reef crest. (After Crossland)

limestone by the sea water. Occasionally, when an especially deep channel has been formed in this manner, living corals may be able to establish themselves once more in this region and, by their subsequent growth, fill up the channel. Finally, nearest the shore will be a second flat, very often formed of coral on its outer side only, the inner part being of land origin, having been formed by the sea which, at the end of its rush across the reef flat, has cut its way into the land.

This explanation of the formation of fringing reefs clearly cannot be applied to barrier reefs or atolls, the outer slopes of which frequently descend to depths of hundreds or thousands of metres. It is the mystery surrounding their mode of formation which has attracted the attention of zoologists and geologists since reefs were first accurately described, and has led to the development of a number of theories which attempt to explain the mystery. The most important of these are discussed below.

Theories of Reef Formation

The first, and most famous, theory concerning the origin of barrier reefs and atolls was put forward by Charles Darwin in 1842. He had observed, during his voyage round the world in H.M.S. *Beagle*, that reef-building corals can exist only in shallow water while, as a geologist, he also knew that the level of the land may vary from time to time. By putting these two facts together he was able to elaborate a theory which, in its simplicity, bears the stamp of genius. He imagined a fringing reef developing, in the manner shown above, off land which was slowly sinking as a result of a general subsidence in that region. As the land sank, the corals would grow upwards and keep pace, more or less, with it, so that the outward edge of the reef would maintain its position in relation to the surface, but the shallow channel between it and the land would become deeper and deeper as the land sank and also broader as, bit by bit, the land was submerged. Finally, a typical barrier reef would be formed, often many kilometres from the new coastline, and with a channel up to 100 m in depth. Here the matter would rest

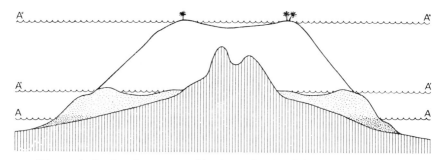

Diagram indicating the manner of barrier reef and atoll formation according to Darwin's views. A–A, sea level around central island with fringing reef. A'–A', the same after some subsidence of the land and upward growth of fringing reef to form a surrounding barrier reef. A"–A", the result of further subsidence; land completely submerged and coral growth now forming an atoll ring with central lagoon

should land remain above water—as in the case of the barrier reefs off Queensland or New Caledonia—but should the land have been a small island which was finally submerged, then all that would finally appear above the surface of the sea would be a ring of coral surrounding a central lagoon, in other words a typical atoll. This process will be made clearer by a study of the figure which shows the various stages in the conversion of fringing reefs into atolls according to Darwin's theory.

At first this theory met with widespread support, notably from an American geologist named Dana, who observed many reefs, but as coral formations began to be studied in more detail and with ever increasing care, it was found that in many regions where corals flourished, so far from there being any evidence of subsidence of the land, there was definite evidence that the land was rising. Moreover, all three types of reefs are sometimes present in the same area, which is incompatible with Darwin's theory according to which the presence of a fringing reef is evidence that the land is stable while the presence of barrier reefs and atolls is evidence of subsidence. Again, Darwin's theory demands that an immense belt of land between the tropics should have been steadily sinking over a long period of time. The theory also involves the presence of numerous islands or high submarine elevations.

Sir John Murray, of whom we have already heard in connection with the investigation of the bottom deposits of the ocean, was the first to bring forward an alternative theory of any general application. The real crux of the problem is how to account for the presence of submarine platforms covered with not more than about 60 m of water, from which alone, barring the subsidence of pre-existing land, corals could grow to the surface. Murray thought these were produced by a raising of the sea bottom. This might take place as a result of a submarine volcanic eruption throwing up a great mound of debris. (These things do happen, islands being occasionally thrown up above the surface of the sea in this way although the light material of which they are composed is usually washed away very quickly.) It might also be brought about by the accumulation

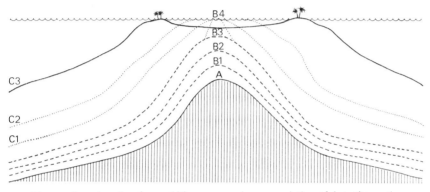

Diagram based on the views of Murray, namely accumulation of deposits on the summit of a submarine elevation followed by settlement of corals and upward growth to the surface

of great banks formed from the skeletons of the planktonic animals and plants in the surface waters which are responsible for the greater part of the deposits on the bottom of the sea.

As soon as a platform was formed in this manner, reef-building corals could be established and begin growing upwards until they reached the surface, as shown here. Clearly those in the centre, at the highest point, would reach the surface first and the preliminary indication of the beginning of an atoll, according to this theory, would be the appearance of a small circular reef flat or patch reef. But the corals in the centre, being unable to grow upwards any further would die, while those on the fringes would grow outwards, extending the flat in all directions. Now, as in the case of the fringing reefs, the force of the sea would throw boulders of coral limestone on to the surface of the flat around the edge and here dry land would begin to form. Gradually the boundaries of the reef would be extended in all directions, more stones and boulders would be thrown up, some of which would be worn down by the action of the weather to form a soil in which seeds carried by the sea, or in floating timber washed ashore, or in soil attached to the feet of sea birds which alighted here, would establish themselves and grow. And so the beginnings of plant life would appear and, by the binding action of their roots, would establish the dry land yet more securely, while their decay after death would provide further, and richer, soil. Meanwhile the centre of the reef flat would become hollowed out to form a lagoon in essentially the same manner as the channel is formed between the fringing reef and the land. The water pouring in between the various islands forming the atoll ring would scour and dissolve away (Murray laid great stress on the latter process) the dead coral in the centre and, with the aid of boring animals and plants, gradually eat out a shallow lagoon. And so the process would go on, the atoll extending outwards like a fairy ring of fungus on grass, the broken fragments or talus falling down the steep slopes outside and furnishing a foundation for the further increase of the reef. In exactly the same manner, Murray thought that a fringing reef could be converted into a barrier reef.

This theory had some points in its favour, notably in that it did not demand any general subsidence of the land over the tropics. It has been supported by other more recent investigators of coral reefs, though often with qualifying statements and additions. The most important criticisms are concerned with the formation of the lagoon, for many think that these are at present being filled up with sediment and not, as Murray's theory demands, gradually increasing by further erosion and solution, and that, in any case, they were originally formed by the scouring action of the sea and by the action of boring animals and plants, rather than by the dissolution on which Murray laid so much stress. It is also very probable that the ring shape of atolls is due to the action of the prevailing winds. All atolls lie within the Trade Wind areas where the wind blows from a constant quarter for the greater part of the year. There is actually an atoll ring off Florida which must have been moulded in this manner. It has not been built by living corals but by deposits of various kinds formed by submerged reefs of very mixed origin.

Clearly a definite proof of the origin of atolls might be expected if borings were made through the coral rock down into the lower layers. If the substance of the atoll was found to be of coral origin to great depths, this would be strong evidence in favour of Darwin's theory, the coral limestone below the normal depth of coral growth having been carried down by the subsidence of the land; if, on the other hand, a surface layer of coral limestone some 60 m in thickness was found resting on a layer of solidified bottom deposits, of volcanic fragments, or similar material, then that would afford evidence that Murray's views are correct. With this end in view expeditions led by Professor Sollas of Oxford and Professor Sir Edgeworth David of Sydney, were sent out under the auspices of the British Association about the end of the last century. The object of these expeditions was to make borings through the atoll of Funafuti, one of the Ellice Islands in the middle of the Pacific. They succeeded after much hard work in boring through the atoll to a depth of 338 m. The core of this bore was carefully preserved and brought to Great Britain where it was examined from end to end by scientists in order to determine its composition and, most important of all, whether it was formed throughout of coral limestone or whether that was succeeded by some different material in the lowest layers. This examination showed that the bore was composed throughout of the one substance, coral lime-stone, and this was at once claimed by the advocates of the Darwin theory as a vindication of their views. But now enters a complication for, without denying that the bore was composed throughout of rock of coral origin, Murray and his adherents stated that this was due to the fact that the bore, instead of going

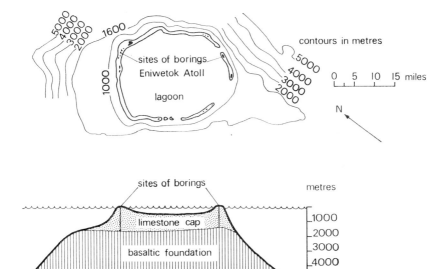

Formation of Eniwetok Atoll. Above, outline of atoll ring with rapidly deepening water around, central lagoon and sites, on marginal islands, of deep borings to determine the nature of underlying rock. Below, section showing results of the borings, namely that the atoll consists of a limestone cap resting on the summit of an extinct volcano. (After Ladd)

through the centre of the reef, had been made through the edge and so had passed through the talus of coral fragments which had broken off the edge of the growing reef and fallen into deep water.

The later views of Professor R. Daly of Harvard now come into the picture. In his 'glacial-control' theory he pointed out that so much water must have been locked away as ice in the polar regions during glacial periods that the surface of the tropical seas must have been lowered, to a depth he estimated of between 50 and 60 m. Exposure would then destroy pre-existing reefs, and the tropical shores they had protected would become fully exposed to erosion by the sea. This led to the formation of extensive platforms which gradually became flooded as the sea level rose at the end of the glacial period. At the same time the water became warmer and corals gradually established themselves on the plat-forms which were eventually covered by up to 60 m of water. The corals would thus grow upwards with the rising water level, not downwards with a subsiding substrate as Darwin postulated.

i

There the matter largely rested until in 1946 the United States began to con-
duct intensive examination of Bikini and adjacent atolls in the former Japanese-
owned Marshall Islands in the North tropical Pacific. This was in relation to the
work of the Atomic Energy Commission which was testing the effects of nuclear
explosions. To find out the depth of the coral mass and the nature of the platform
on which it rested, borings were made there and, as shown in the figure, on the
neighbouring Eniwetok Atoll. The two made there passed through 1265 and
1405 m of coral rock before each encountered basalt. In other words the atoll
was revealed as an immense cap of limestone, all of coral formation, resting on
the summit of an extinct submarine volcano which itself projected some 3 km
above the floor of the ocean. So, in this region certainly, and there is much
evidence to the same effect in the West Indies, Darwin has been proved correct.
Corals could not have grown upwards from these great depths; the surface on
which they rested must have sunk but never so fast that the corals could not have
maintained contact, if not with the surface of the sea at least with the shallow
illuminated waters in which alone they can live. When they were able to break
the surface the steady force of seas driven by the Trade Winds could have been
responsible for the moulding of the atoll ring.

Meanwhile there has also been the discovery in the Pacific of numerous isolated 'sea mounts', flat-topped volcanic cones on the summit of which, were
these near enough to the surface and in sufficiently warm water, corals could be
established and grow towards the surface. There seems little doubt that modern
atolls were built up on the summits of such mounts which gradually sank as a
result of the growing weight of coral. Moreover the views of Daly have been
enlarged and extended. The five glacial periods now known to have occurred
during the past million years must all have been responsible for lowering of the
sea level with return of this during the succeeding interglacial period. At maximum, sea level must have dropped to some 100 m below its present level. There
is clear evidence of a somewhat intermittent rise in level over the past perhaps
10,000 years since the last Ice Age. In other words all existing reefs are quite
recent though often growing on the surface of pre-existing reefs belonging to the
Pleistocene period. During the Ice Ages reefs must have retreated both in depth
and towards the equator, then moving upward and farther to the north and to
the south as the sea level rose.

Chapter 8
COLOUR AND LUMINESCENCE

In this short chapter we shall consider two of the most interesting properties of marine animals, their colour and the strange power which many of them possess of producing light so that they appear luminescent in darkness. Though colour and luminescence are not directly connected they have this at least in common, that the production of the former is greatly influenced by light whereas the latter is concerned with the production of light.

Colour

In the sea there are animals with as great a wealth and variety of colour as any that are found on land. This is a fact which people who live in temperate climes are liable to overlook because our fishes usually have none of the brilliancy of colour of those from tropical waters which, however, may be viewed in the now numerous coloured underwater photographs of coral reefs. The most beautiful of our marine creatures are anemones which often lie concealed in rock pools or contracted on the sides of rocks while the tide is out, or else small shrimps and sea slugs which are known only to the naturalist with the knowledge and enthusiasm to search for them.

The remarkable property of colour change—whereby an animal can alter its colour to a greater or less extent, usually in order to tone with a different background—is possessed by a far greater number of marine than of terrestrial animals. It is especially well developed in crustaceans, in squids and cuttlefish and in fish. Amongst the first of these the most striking example is that of the little Æsop prawns (*Hippolyte*). These creatures, less than 2·5 cm long, are very common on seaweed in rock pools and on rocky coasts although very difficult to discover owing to the exactness with which they tone with the background.

Beginning with the birth of the prawn; when the young hatch out they are colourless and almost transparent and drift about in the sea for some time, being finally carried inshore where, for the sake of the food and security afforded them, they cling to the first piece of seaweed they meet. They now begin to develop colour, a process which takes place quickly and always results in the animal assuming the exact shade of colour of the weed on which it happens to have settled. If this is a red weed the prawn becomes red, if a brown weed from the intertidal zone, it becomes brown, or if a green weed in the rock pools near high-water mark, it becomes green. The colour change begins gradually but at the end of a week the match in colour is usually perfect and the prawn almost impossible to detect.

If the Æsop prawn is driven from its home by unfavourable conditions, such as strong tides or currents, it searches for a new home of the same colour, but should this be impossible (as it may easily be made in the laboratory by placing collected prawns on weeds of different colour from themselves) then the prawn changes its colour to that of the new background and at the end of a week will have developed a new colour scheme just as exact an imitation of its surroundings as was the first. Quite independent of these enforced changes of colour due to change of surroundings are the nightly changes of colour which occur with unfailing regularity. Whatever colour prawns may have by day, they *all* change to a beautiful transparent blue by night, the alteration in colour taking place quickly shortly after nightfall.

Great numbers of minute pigment cells scattered over the surface of the body are responsible for the colour of the Æsop prawn. Each of these cells consists of a central bag with many fine extensions. Within the central region there may be three pigments, red, yellow and blue, and the colour of the animal is controlled by the extent to which these pigments are spread through the branches, thus covering the maximum area, or are withdrawn centrally where they are reduced in area to a minute point and so hardly affect the general colour of the animal. The three colours may be associated in different ways; thus the red pigment alone occupies the branches when the animal is red, the green colour is produced when the red pigment is withdrawn into the central bag and the yellow and blue pigments together spread into the branches. The nocturnal blue colour is produced when the red and yellow pigments are both withdrawn and only the blue one extends into the branches of the pigment cells.

What is the mechanism which controls these remarkable and exact changes in colour? It appears that the pigment cells are influenced by the light which acts either by way of the eyes or directly on the pigment cells. Different coloured light affects the hormones controlling the flow of pigment in different ways, and ensures that the colour that is produced shall tone with the background of the animal. The supreme importance of these elaborate mechanisms for colour change in these little animals is clear. There can be no doubt that the protection provided by this cloak of invisibility is very great because they are not very active animals and enemies are numerous in the shore waters where they live.

Anyone who has ever examined a living squid, or one that has only very recently died, cannot fail to have been impressed by the wonderful play of colour which sweeps over the torpedo-shaped body which blushes with delicate colour, and then as quickly blanches. In this case also the colour is controlled by tiny pigment cells, or chromatophores, in the skin but these are of a different nature from those of the Æsop prawn. Each contains one colour only and this is not spread into branches or withdrawn according to need. Each pigment cell consists of an elastic bag to which are attached radiating muscles. The pull of these dilates the cell which immediately exposes a surface of colour, with accompanying reduction in this when the muscles relax. This colour change due to muscular action is extremely rapid and waves of colour may spread over the surface of cuttlefish or squids. By suitable colour change the animals may adapt themselves to the background but they may also distract the attention of their prey or of their enemies. Colour change plays a notable part in their behaviour.

Many fish have the power of colour change. This is especially true of the flat-fishes, such as the sole, turbot, plaice or flounder, which live on the sea bottom and are liable to be attacked from above. If they are examined it will be seen that, whereas the underside of the body which lies on the bottom is unpigmented, the upper surface is mottled grey and brown so as to provide an almost perfect match with the background of mud, sand, or gravel. Here again pigment cells in the skin are responsible. These are of several varieties, some being black and others orange or yellow, while there are also bundles of crystalline needles which act as reflectors. The pigment can be extended or withdrawn, contracting to tiny points or ramifying into the branches and on the relative degree to which the different pigments are expanded or contracted depends the colour pattern of the fish. It is perfectly easy to experiment with flat-fish on different coloured back-grounds. In this case light appears to be the deciding influence in colour change, though it never acts directly on the pigment cells, always by way first of the eyes and then the nerves. A blind fish is not able to change colour. The lack of pig-ment from the underside of the body is apparently due to the normal absence of illumination on this side because if the fish are placed in glass-bottomed tanks illuminated from beneath, they develop colour on the under, as well as the upper, surface.

Many marine animals are coloured in such a way as to blend with their sur-roundings, but do not possess the power of changing colour. This is true of many of the sea slugs, such as the common sea lemon (*Archidoris pseudoargus*) which lives on rocks and is yellow with mottlings of red and other colours, closely resembling the rock with its encrusting sponges, sea squirts, and red coralline seaweed. An allied slug, the red *Archidoris flammea*, always lives on red sponges; there is a small green sea slug named *Elysia* which always lives on green weed, and this list could be extended almost indefinitely. Many crustaceans are also difficult to detect in their natural surroundings and in this group, it may be remembered, there are many instances of 'masking', the animals draping them-selves in seaweed, sponge, or hydroids the better to disguise themselves.

In other cases the often vivid colouring of animals like anemones, many worms, starfish, or sea squirts, appears to have no value to the owner, at any rate from the standpoint of protection.

There are well defined differences between the colours of animals inhabiting various depths in the ocean, as indicated in Plate 40. Fish from the surface of the sea in tropical regions are usually sky-blue, whereas deep-sea fish are always darkly coloured, black, brown, violet or red. During the cruise of the *Michael Sars*, hauls made in the Sargasso Sea revealed the presence near the surface of blue flying fish, others of a silvery colour and the transparent young stages of such fish as the eel; at depths of 300 m fish with silvery sides and brownish backs were captured, while below 500 m only black fishes and red prawns were found.

It is noteworthy that these black fishes and red prawns only live in the regions where no or very little light can penetrate, living nearer to the surface in the

Norwegian seas than in the tropics owing to the reduced extent to which light penetrates in the former. Another point of interest is that different species of the same genus, or even different varieties of the same species, which live at different depths are variously coloured, those from the greatest depths having the most intense and darkest pigmentation. The fishes from the abyssal region where there is no trace of light whatever appear to be usually brown, blue or violet, though what part pigmentation can play in the lives of animals always surrounded by impenetrable darkness it is impossible to say.

The connection between light intensity and colour in marine animals is difficult to explain; in some cases the colour produced may be protective but this is probably only the *result*—possibly quite an incidental one—and not the cause of the colour produced. We still know all too little about both the function and often even the chemical nature of the colours of marine invertebrates. The composition of those that are understood chemically varies very greatly. Some pigments are produced by the animal itself; others are derived from their food. They are said to be respectively endogenous and exogenous. A good example of the latter is provided by the common dog-whelk (*Nucella*). The shells of the majority are white but some are banded with deep brown and the animals forming them are known to feed on mussels instead of the acorn barnacles which are the more usual food. Dog-whelks in exposed situations have sometimes a yellow shell the cause of which, probably endogenous, is unknown.

Luminescence

Although many land animals, such as the glow-worm and firefly, are able to produce light so that they glow at night, this power of phosphorescence, or 'bio-luminescence' as it is better called to distinguish it from light produced by phosphorus, is very much more widespread amongst marine animals. It is so common that it has aroused interest from the earliest times and many theories have been advanced in the attempt to explain the mystery of its production. For instance the explorer Franklin thought that phosphorescence in the sea was due to friction between the salts producing tiny electric sparks. This and other views are now of merely historic interest for, though the power of bio-luminescence is possessed by most diverse animals, and is displayed in many different ways, the actual production of light is the result of a chemical process which is identical in almost all cases. As long ago as 1667 the great chemist Robert Boyle found that bio-luminescence could only take place in the presence of air while, owing to the work very largely of an American physiologist, E. Newton Harvey, we now know that the light is produced as a result of the oxidation of a substance called luciferin. When this combines with oxygen, light is produced, the luminescent reaction being probably represented by the following equation:

$$\text{luciferin} + \text{oxygen} = \text{oxy-luciferin} + \text{water} (+ \text{light}).$$

The reaction is essentially the same as any other oxidation, for example that which occurs when a candle is burnt, but the energy produced takes the exclusive form of *light*. This is one of the most striking properties of this animal light—

it is *cold*, being unaccompanied by heat of any kind and is consequently more efficient than any of the forms of light which we can artificially produce, in all of which there is an invariable waste of energy owing to the accompanying production of heat.

But luciferin is always accompanied by a second substance known as luciferase. This is an enzyme—belonging to the same class of substance as the enzymes which enable us to digest our food—and takes no direct part in the production of light but acts as a kind of chemical lubricant assisting in the union of the luciferin with the oxygen. Both luciferin and luciferase can be isolated from the animals that produce them so that it is possible to produce animal light in the laboratory and to experiment with it in various ways. When the two substances are placed in a test-tube together, there is a momentary glow near the surface, where alone

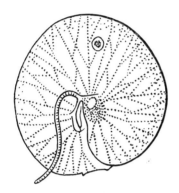

Noctiluca miliaris, a flagellate protozoan which is luminescent. (Greatly enlarged)

there is available oxygen, but if the tube is shaken further oxygen is admitted and more light is produced either as a glow or as a sudden flash of light.

The oxy-luciferin can be deprived of the oxygen with which it has united—can be 'reduced' in chemical phraseology—and so converted back into luciferin and can again be used, when luciferase and oxygen are present, to produce more light. This process must apparently also go on in nature, there being a continuous change in the light organs of some animals from luciferin to oxy-luciferin with the production of light and then back again.

Certain bacteria are the lowest form of life to produce light. They are responsible for the light which appears on dead fish, and they occasionally invade living animals, giving them a luminous appearance. Some cephalopod molluscs and bony fishes normally harbour such bacteria in special pouches from which a continuous light is produced. Of the single-celled animals, the Protozoa, two, namely *Noctiluca miliaris* and *Pyrocystis noctiluca*, have exceptional powers of luminescence. They have no special organs for light production but contain innumerable minute granules which are responsible for the luminescence. The presence of these animals in countless numbers in the sea gives rise to that general phosphorescence, or 'burning of the sea' as it was called by earlier observers, which has probably been noticed by everyone who has had much experience of the sea by night. *Noctiluca* is especially abundant in the plankton during late summer and autumn and in this season the prow of a boat cuts a silvery line through the water—occasionally of such brilliance that even in the dead of night a newspaper can be read by the light it casts. *Noctiluca* itself is pinkish in colour and may be thrown upon the shore in such numbers as to form a coloured layer along the beach, the shore water resembling 'thick tomato soup'.

According to Murray and Hjort, the Norwegian fishermen distinguish between 'dead phosphorescence' and 'fish phosphorescence', the former resembling 'the stars in a clear sky, myriads of minute nearly invisible points emitting a scintil-lating light, now increasing, now decreasing, in intensity' and being produced by *Noctiluca* and similar Protozoa. The second type, the result of rushing move-ments of large fish or squids which cause the luminescent animals momentarily to glow, is more like a dull glow of light which suddenly lights up and then dies out completely.

Many jellyfish have the power of general luminescence, appearing as round balls of white fire in the sea. Those seen and described by travellers in the warmer seas are often of great size, and the late Professor Herdman described how, when at anchor on the pearl banks of the Gulf of Manaar, 'in an intensely dark night, I saw the black sea around us in all directions lit up by an innumerable assemblage of what looked like globes of fire, waxing and waning in brightness, all simul-taneously glowing and then fading away into darkness, and after a few seconds lighting up once more. This periodic display continued for about an hour and then disappeared.' In British seas the most vividly luminescent jellyfish is the rounded *Pelagia noctiluca* (Plate 44) which is, however, much more abundant in the Mediterranean. The method of light production is probably similar to that of *Noctiluca* only on a much larger scale; there are no special light organs, but a stimulus of any kind, such as that caused by the passage of another animal, causes the whole animal to glow with light. Another animal related to the jelly-fishes and even more intensely luminiscent is the sea pen, *Pennatula phosphorea* (Plate 44), which lives in mud off the west of Scotland and Scandinavia. When brought to the surface the slightest touch on any part will cause that region to light up in the darkness, the light then spreading from branch to branch until the whole is aglow. The large sea pen (*Funiculina quadrangularis*) which may be 1·8 m long, is especially luminescent along the main stem which, if gently touched, glows with light which travels up and down like a flickering flame.

The mechanism of light production in certain hydroid medusae is unique. It consists of a luciferin without luciferase but which luminesces in the presence of calcium, for the presence of which it represents a test of altogether exceptional delicacy. This appears in the jellyfish *Aequorea* and in this instance is known as 'aequorin'.

The most luminescent worm is, peculiarly enough, the obscure *Chætopterus* which spends its whole life within a parchment-like tube buried in mud. This animal produces a luminous substance mixed with mucus from many parts of the body. The light in this case is usually violet or bluish green, and a similar colour is given by many other small marine worms. In some of these, such as the delicate *Odontosyllis* from Bermuda, the production of light is concerned with reproduction, light being produced only during the reproductive season when the animals swarm in the sea for mating.

Coming to the crustaceans we find much more highly developed powers of light production, definite light organs or photophores, some of them very simple but others of great complexity, being found in many of these animals. In some of the simplest cases, luminous slime is produced by little glands above the mouth, the substance discharged glowing with a yellow light. A few of the ubiquitous copepods of the plankton are luminescent; when disturbed they throw off a cloud of luminescence produced by glands spread over the surface of the body. Some prawns and the shrimp-like euphausiids—large planktonic crustaceans

Chaetopterus variopedatus, a luminescent tube-worm as it appears in the dark after removal from the tube. (Natural size)

which are the major food of whales—have very complex light organs or photophores, so intricate that they were originally thought to be additional eyes! There are usually ten light organs, a pair behind the eyes, two pairs on the side of the body and four on the underside of the tail, and each of these consists of a layer of light-producing cells, behind which is a reflector backed by a layer of pigment which prevents any of the light from being wasted by passing into the body, while in front is a lens which focuses the light. Into the light-producing area passes a nerve which controls it—for the luminescence of these animals is under direct control and the light organs can be turned off and on at will like an electric torch. The organs are also well supplied with blood which supplies the oxygen necessary for the production of light, from the blood stream. A typical light organ of this kind is shown in the figure on page 128.

In the chapter on boring life we discussed the piddock (*Pholas dactylus*), a bi-valve mollusc which bores its way into rock with only the tips of its siphons pro-jecting into the water. This creature, strangely enough, is one of the most luminescent of marine animals, a fact which has been known from very early times. Light is produced within minute glands present in five areas of the body. The light is a greenish blue and very powerful. Species of squids and cuttlefish may all have highly elaborate, luminous organs, especially the squids which swim near the surface and sometimes those from deeper water (Plate 44). Though occasionally the light organs are simple, they usually consist of a light-producing area backed by reflectors and a protecting coat of pigment, very similar to those we have already described for the crustaceans, the light being focused by a lens and the whole protected by a cornea, just as in the eye. Some deep-sea squids have over twenty light organs in various parts of the body; most of these throw a white light but a few are deep blue, two near the eyes are usually sky-blue while two organs on the under end of the body—like rear lights—are

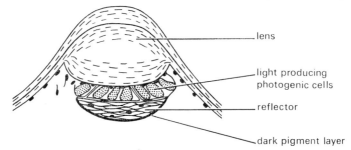

Light-producing organ (photophore) of a deep-water crustacean, *Sergestes*.
(After Terao)

most appropriately red. The different colours are probably the result of coloured screens in front of the light organs.

The interesting tunicate named *Pyrosoma*, a glass-like transparent colony of small animals which unite one with another to form a hollow cylinder, one end of which is closed, is found drifting about near the surface of the sea and forms a conspicuous member of the plankton in such regions as the Mediterranean and the tropical Indian Ocean. Each member of the colony has two light-producing glands and the light they give out is said to be red in some and blue in others, the whole colony, when stimulated, blazing with thousands of these tiny points of light. Moseley, one of the naturalists on the *Challenger*, states that, after a *Pyrosoma* more than 1·2 m long had been captured, 'I wrote my name with my finger on the surface of the giant Pyrosoma as it lay on deck in a tub at night, and my name came out in a few seconds in letters of fire.'

At one time it was thought that luminous fishes were especially common in the deepest seas, regions of utter darkness, the light being used to assist them in their search for food. This is now known to be incorrect for, though the fish living on the surface, such as the herring or mackerel, never display luminescence, yet this power is commonest in fish living at moderate depths, on the bottom or in the intermediate regions. Light organs are apparently commonest in fish which live in the mesopelagic zone between 200 and 1000 m where some measure of light penetrates. But there are notable exceptions, for example, a remarkable fish called *Harpodon* which lives in rivers and estuaries in India and, when caught, displays the most vivid luminescence, and another called *Photoblepharon* found in pools of fresh water in quarries and the craters of extinct volcanoes in Malaya. The actual organs vary a great deal in both structure and position. They may be mere pits and channels for the production of a luminous slime, such as are found in some of the deep-sea Macruridæ, or they may be complex organs like the natural lamps we have described in some of the crustaceans and squids. They

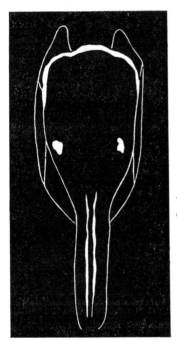

The rock-boring piddock, *Pholas dactylus*, showing luminescent regions as seen in the dark. (Natural size)

may occur on almost any part of the animal, and are often arranged in rows along the sides of the body though they are frequently found in the head region, as in the case of the deep-sea anglers with their waving lamps for the probable attraction of prey. One very unusual fish has no eyes but instead a large light organ under the frontal bones within the skull. A remarkable angler fish brought up from a depth of some 3500 m by the *Galathea* expedition, and suitably named *Galatheathauma*, has a forked light organ within the mouth.

The reason for the production of light in these different marine animals is in most cases a complete mystery. We can only hazard a guess as to what part it plays in the life of the animal. Thus it is difficult to see what use luminescence can be to bacteria, *Noctiluca*, jellyfish, or the sea pens; it is probably merely a by-product of the normal activities of these creatures. The case of the concealed animals, such as the piddock or the tube-worm (*Chætopterus*) presents almost equal difficulties though in the latter it may be that the light serves to attract to the burrow the tiny organisms on which it feeds. In other cases the light may help members of the same species to recognize one another or assist the females in attracting the males during the mating season. It appears that luminous

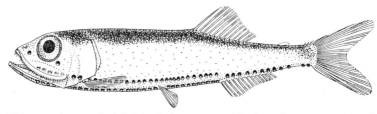

Mid-water (mesopelagic) fish of deep waters, *Vinciguerria attenuata* (4 to 5 cm long), showing enlarged eyes and photophores on the head and in rows along the underside of the body

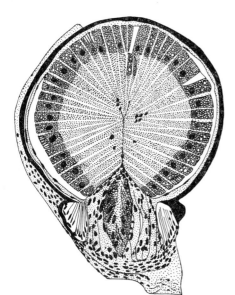

Section showing the complex structure of the photophores in *Vinciguerria* (Greatly enlarged)

secretion may be discharged into the water or photophores turned on to distract a pursuing enemy, while in other cases the sudden flashing of light from the more complicated light organs may perhaps protect the animal by giving warning that it is harmful or distasteful as food, in much the same way as the brilliant colours of some terrestrial animals are said to afford warning of the unpleasantness of their owners. There is now evidence that the downward directed photophores in mesopelagic fishes such as *Vinciguerria* tend to make them invisible from below when they would otherwise be visible against the faint light coming from above. Of course it is possible that there is truth in the old belief that the light organs of the abyssal animals which look so much like little lanterns are actually used for this very purpose and help them in their search for food. But we can be certain of nothing and the precise biological significance of animal light is, and seems likely to remain for some time to come, as deeply mysterious as it is fascinating.

We must not leave this subject without referring to the practical use to which the presence of luminescence is put by the natives of the Banda Islands who employ the luminous organs of a fish as bait in fishing and with a success which justifies their enterprise.

Chapter 9
FEEDING OF MARINE ANIMALS

Food is the first necessity of life, and nowhere is the struggle to obtain it keener than in the sea; every possible source of food has been exploited by one type of animal or another. Hence the study of the multitudinous devices for obtaining food of one sort or another, which have been evolved as a result of the keen struggle for existence, forms one of the most fascinating branches of marine biology.

Food in the sea consists of inorganic and organic matter. The former comprises dissolved gases and salts such as carbon dioxide and the nutrient salts, nitrates, phosphates and sulphates in particular, which together form the food of plants—both the microscopic members of the plant plankton and the coastal seaweeds. From this inorganic material they form first carbohydrates, that is, sugars and starch, and then fats and proteins. This initial building up of organic matter, which can take place only in the presence of sunlight from which the necessary energy is drawn, constitutes primary production. The animals which feed on the organic matter so formed, namely the herbivorous members of the plankton and many members of the bottom fauna in shallow water, together with the relatively few which feed on seaweeds, represent secondary production. On them in turn prey the carnivores which may seize living prey or scavenge amongst the moribund or dead. They may be said to form tertiary production and the matter may not end there but in form of still larger carnivores, such as sharks, toothed whales and seals, all of which feed on smaller carnivores such as fish and squids. At each stage, from plant to herbivore, from herbivore to carnivore (to take matters no further) there is a loss of up to 90% so that tertiary production (e.g. the quantity of food fishes which are all carnivores) may be little more than one hundredth of primary production.

In addition to the true herbivores and carnivores there are animals which feed indiscriminately upon many kinds of food; but owing to our lack of knowledge concerning the food of many marine animals, it is perhaps better to classify them according to the method by which they feed rather than by the particular substances which they eat. Thus we may divide animals into those which possess the means for feeding on fine particles, on large masses or on living prey, and finally for sucking in fluid food. To these must be added the parasites which prey upon other animals and those remarkable cases of intimate union between two animals or between an animal and a plant, the two becoming entirely or partly dependent one upon the other, a condition known as 'symbiosis', from two Greek words meaning 'together' and 'life'.

Invertebrates Living on Finely Divided Food

The animals which feed on microscopic plants and animals or on fine particles in suspension in the sea are usually epifaunal animals attached to the bottom or members of the infauna which burrow or bore into it, or else they are small animals of the plankton. Many of the mechanisms with which they are provided for collecting their finely divided food are extremely complicated, and the mode of feeding in many of these animals is the most elaborate found in the animal kingdom.

A very large number of these animals create a stream of water by means of rapidly moving ciliary or flagellar hairs, the food particles being sieved out and swallowed. The simplest mechanism of this type is found among the sponges which are honeycombed with a series of fine canals, through which water passes to be later expelled through a large central opening or 'osculum', all food particles in the water having been engulfed in the flagellated cells during the process. Sedentary tube-worms, such as sabellids and serpulids (p. 27), bear round their head end a crown of foliaceous tentacles, often of great delicacy and beauty. This normally projects freely in the water but is withdrawn in a flash on the approach of enemies. The tentacles are covered with cilia and also with a sticky mucus in which fine particles are entangled, being later carried to the mouth by currents produced by the beating of the cilia. But it is the bivalve molluscs, such as the oyster and mussel, which have carried this method of feeding to the greatest degree of perfection. In these animals the gills are enormously developed and form two deep membranes running down each side of the body, as shown on page 230. Each membrane consists of many parallel filaments united from place to place and so forming a fine meshwork. Everywhere the gill surface is ciliated and a powerful current of water is drawn in on the underside of the animal (see lower arrow); this then passes through the gill to leave on the hinder surface (upper arrow). All particles in this water current are retained on the surface of the gill, where other tracts of cilia carry them, together with the mucus in which they become entangled, forward towards the mouth. They do so either along the axis of the gill, where the two membranes are attached to the body, or else along the free margin of each membrane where there is a fine groove. Before reaching the mouth this potential food—to the extent at any rate that it consists of finely divided plants of the plankton—is carried between pairs of triangular flaps which guard each side of the mouth. These lips or 'palps' are ridged on their inner surfaces and are even more elaborately ciliated than the gills. They constitute a very complicated but remarkably efficient mechanism for sorting out the particles, the larger ones being rejected. In oysters these accumulate just within the margin of the shell where they are conveyed by still other tracts of cilia. From time to time the shell closes, owing to sudden contractions of the centrally placed adductor muscle and this waste matter is expelled. This highly efficient feeding mechanism represents, it will be observed, an immense development of the gills. The respiratory current is greatly increased (far beyond the animal's need for oxygen), while cilia originally concerned with cleansing the surface of the gill take on the new functions of food collection and its transport to the mouth.

The sea squirts feed in a somewhat similar manner, water being drawn in through one opening and expelled by the other after having been sieved through a delicate latticework. So fine is this lattice that, in a medium-sized sea squirt, it has been estimated that there are almost 200,000 openings and about double that number of rows of cilia, the beating of which creates the current. Here again the food particles are entangled in a sticky mucus and carried to the mouth by special tracts of cilia. Near relatives to the sessile sea squirts are the tiny floating appendicularians, which form part of the plankton. They feed in truly amazing manner. They do not catch the food directly but form an elaborate gelatinous 'house' much larger than themselves. As shown in the figure, this house is really a complicated apparatus for straining sea water. As a result of the beating of the

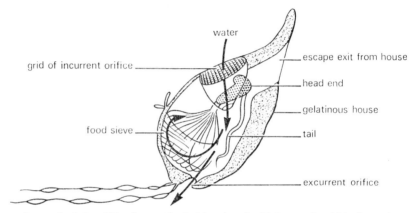

Appendicularian *Oikopleura*, stippled head end with long tail, within 'house'. The direction of water flow through this produced by movements of the tail shown by arrows. Full description in text. (× 5)

tail of the occupant, water is drawn in on the upper side of the house through a pair of openings guarded by a fine network and the extremely fine particles still remaining in suspension are collected by a very complicated sieve within the house. The food is then drawn into the mouth of the animal, again by the aid of cilia, while the strained water passes out. The house soon becomes useless on account of accumulation of particles which effectively clog it. It is then abandoned by the animal—after perhaps being in use for only a few hours—and a new one in all its intricate detail constructed in half an hour or so! Other members of the plankton which feed exclusively by means of cilia include some of the delicate pteropods or sea butterflies.

Many of the smaller crustaceans have feathery limbs for collecting finely divided food. The common barnacles (Plate 11) are provided with six pairs of delicate limbs fringed with fine bristles, which are alternately shot out from the shell and withdrawn, sweeping through the water like a casting net during the latter phase. The process can easily be observed by placing acorn barnacles collected from the rocks in a basin of sea water and cannot fail to be admired for the grace and precision of the movements. Barnacles may filter fine particles from the water or actually seize larger objects, or they may do both. Many of the smaller planktonic Crustacea, such as the ubiquitous copepods, by the movements of their swimming appendages create currents of water in which food particles are carried between the appendages and then forward along the underside of the body. There the food is filtered out by a mesh-work of hairs or setae and is then pushed into the mouth. Some of them feed exclusively on diatoms, others on smaller copepods, and others again on a mixture of the two.

The small sea gherkins (such as *Cucumaria* and *Thyone*) which live in holes and cracks in the rocks and have a ring of branching tentacles around their mouths, have an original manner of feeding. The tentacles are usually projected

Holothurian or sea cucumber, *Cucumaria*, feeding by means of the branched tentacles around the mouth. (Natural size)

out of the rock and, as they are covered with mucus, food quickly collects upon them. There are no cilia to carry the food to the mouth; instead the tentacles are deliberately curved inwards and pushed to their fullest extent into the mouth and then slowly withdrawn, the food being scraped off by means of a pair of short Y-shaped tentacles at the side of the mouth. One tentacle after another is pushed into the mouth in regular sequence and sucked clean—exactly like a child sucking jam off its fingers! A marine snail called *Vermetus* feeds in a unique manner. The shell, largely uncoiled, is cemented to the substrate. With the consequent loss of movement the foot is reduced, but the mucus used in other snails to lubricate movements is here employed to entangle food. Plankton and suspended particles are caught in extruded mucus strings which from time to time are pulled in and pushed into the mouth.

Burrowers and Scrapers

A great many animals (such as worms and sea cucumbers) live in mud, spending their days slowly ploughing through it and swallowing large quantities from which they extract such nourishment as it contains, much as the earth-worms do on land. The burrowing sea urchins feed in somewhat the same manner, but they pick up particles by means of special grasping tube-feet, which surround the mouth. Other animals of which shipworms (p. 95) are examples, obtain their food by boring; while others again scrape off the various plants and animals which form a crust over the surfaces of the rocks. Of the latter type of feeders, the common sea urchins are examples; they hold on to the rock with their tube-feet and bite off the food by means of five long teeth, which are supported in an intricate skeleton shaped like a lantern and named after Aristotle who first described it. The contained muscles force the teeth downwards so that they converge beneath the mouth, biting off a rounded piece of food, which is automatically swallowed.

Other animals which scrape their food from the surface of stones are the common shore snails, such as the periwinkle and the limpet, which crawl about by

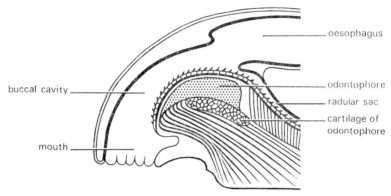

Gastropod mollusc, diagram of radular apparatus, shown withdrawn. Description in text

means of the big muscular foot. In common with other members of the snail family they possess a very characteristic feeding apparatus consisting of a long horny ribbon, made up of many rows of fine teeth, and known as the 'radula' or lingual ribbon. This is supported by a strengthening framework over which it is drawn backwards and forwards like a rope over a pulley. The whole mechanism can be withdrawn into the mouth when the animal is not feeding, but when in use is pushed out against the food which it rasps away by continuous backward and forward movements, scrapings being passed back into the gut as by a conveyor belt. The radula is constantly being worn away, but is just as steadily replaced, new material being continuously added to the hind end of the ribbon. This varies very much in different snails depending on their particular type of food: thus in the animals we have been discussing it is broad, the better for scraping over wide surfaces, while in the carnivorous snails, which we shall discuss later, it is much narrower with fewer but larger teeth. Jaws may also be present especially in herbivorous snails.

Predacious Carnivores

Great numbers of animals in the sea prey upon other animals. All the anemones and jellyfish, in spite of their delicacy and great beauty, are really voracious carnivores. They seize their prey, which may be anything from worms to small fish, by means of their tentacles. These are armed with batteries of minute 'nettle cells', each of which contains a tiny bag filled with a poisonous fluid and is drawn out into a fine whip-like process which lies coiled within the bag. When they are touched by any animal these nettle cells explode and the hollow thread, which is

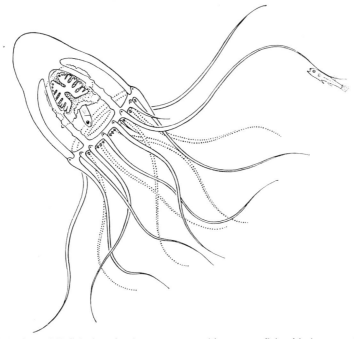

A medusan jellyfish, *Leuckartiara octona*, catching young fish with the tentacles which transfer them to the mouth within the bell. (×2)

usually barbed, is violently everted into the body of the prey. The effect of the poison is to cause paralysis which is usually followed by death. The prey is then pushed into the mouth and so into the stomach, where it is digested with remarkable speed. Many jellyfish will seize and swallow animals larger than themselves, seizing young fish with their delicate tentacles and playing them like an angler. Starfish and brittlestars crawl about on rocks or in the sand and devour any suitable animal which they can find, the former seizing them with their tube-feet, and the latter wrapping them round with their very active arms. In both cases small prey is carried to the mouth and swallowed, but starfish are also able to feed on animals much larger than themselves by the convenient process of protruding the stomach over the body of an animal which cannot be swallowed.

They consume great numbers of bivalves over which they hunch their bodies and proceed to pull the shell valves apart (at a recorded rate, in some, of 1 mm a minute) by means of their tube-feet. As soon as the shell gapes adequately the stomach wall is protruded and almost 'flows' over the tissues within. It probably first poisons and then certainly digests the soft body of the bivalve. It is this habit which makes the starfish a major pest on oyster and mussel beds and the large crown-of-thorns on corals (p. 112).

Many worms are carnivorous, seizing their food by means of horny jaws and usually swallowing it whole. Whelks and their relatives live on carrion, often dead or dying bivalves; their mouths are situated at the end of a long proboscis

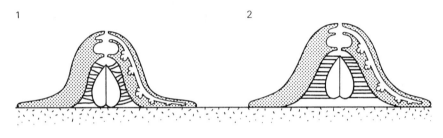

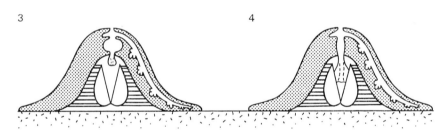

Diagram showing how a starfish feeds on a bivalve. 1 Arches over bivalve with tube-feet attached to valves. 2 Steady pull. 3 Resistance of closing muscles overcome and valves gape. 4 Stomach protruded through mouth and proceeds to digest body of the bivalve

which can be extruded for a considerable distance, and in this manner the inside of a mussel can be cleaned out, the flesh being scraped into the mouth by the radula. Other marine snails such as the oyster drills, *Ocenebra* (*Murex*) and *Urosalpinx*, attack living bivalves, penetrating the shell by a combination of chemical action from a gland in the foot and mechanical penetration by the radula. As soon as the opening is made the long proboscis with which all these carnivorous snails are equipped is pushed into the soft flesh which is very quickly consumed. The American whelk (*Busycon*) enters the shells of oysters by moving on to them and waiting until the oyster opens when, by a sudden twisting movement, it drives the edge of its own shell in between the open valves, which are thus held apart sufficiently to allow the proboscis to be inserted. Other snails grip the shell of their prey with the broad foot and force the edges together, cracking off pieces near the edge until a hole is made large enough to allow the proboscis to enter.

The tropical cone shells (*Conus* spp.) are highly active predacious carnivores, literally pouncing on their living prey. The radula is reduced to a series of separate hollow teeth each charged with poison produced by the modified salivary gland. The harpoon-like tooth is literally shot into the body of the prey which, in different cones, may be fish, worms or other snails. The prey is then swallowed whole into the extended proboscis. The poison in several species is lethal to man.

The sea slugs frequently browse on stationary animals: thus the sea lemon, *Archidoris*, spends its life scraping sponge into its mouth with its broad radula;

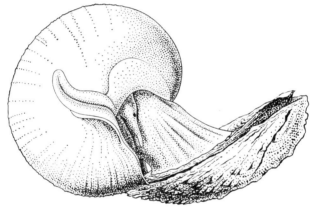

Whelk, *Busycon*, opening an oyster, holding this by the foot and inserting the edge of its shell between the valves so prizing these apart. (After Colton)

the beautifully coloured *Æolidia* feeds on anemones, being completely unaffected by the nettle cells. It is a very remarkable fact that these nettle cells, after they have been swallowed, are transported into projections on the backs of such æolids (Plate 12) where they establish themselves and in which they are always found, so that in the past naturalists thought they actually developed there. Sea slugs are not attacked because they are distasteful; in æolids the harboured nettle cells assist in such protection. One little sea slug called *Calma* feeds exclusively on the eggs and embryos of shore fishes, which are laid on stones and shells. When feeding, its face fits like a hood over the egg, which is slit open by the narrow, saw-like radula and the contents swallowed. There is another type of carnivorous marine snail, the boat-shell (*Scaphander*) which burrows in sand, swallowing small bivalves whole and then crushing them in a gizzard lined with limy or horny plates and worked by powerful muscles.

The active pelagic squids pursue shoals of fish near the surface. The octopus and cuttlefish lie concealed, the one in crevices among rocks and the other in the sand, and dart out upon their prey, usually fish or crabs, which they seize by means of their tentacles armed with suckers. They are particularly careful when attacking crabs to seize them from behind so that these are unable to defend themselves by means of their claws, which are gripped firmly by the suckers and held away from the body of the attacker. The mouth of all these cephalopods is in the centre of the arms and possesses a pair of extremely powerful horny jaws, shaped like the beak of a parrot, and also a small radula. After the prey has been seized it is bitten into by the jaws and a poison injected in which quickly kills it. In the less active octopods the flesh is dissolved away by means of re-gurgitated digestive juices and the fluid sucked in through the mouth. In this manner the shell of a crab is entirely cleaned out and then discarded.

The crab and lobster family are mainly scavengers, feeding on whatever they can obtain, dead or alive, plant or animal. They seize their food by means of their powerful pincer-like claws in which it is first crushed and then pushed towards the mouth which is guarded by a whole series of jaws and other appendages by means of which the food is torn up and shredded out until it can be easily swallowed. The stomach of these animals is lined with chitin and also possesses three teeth which are worked by muscles in such a way that all three come to-gether in the centre of the stomach, breaking up the food still further. It is then acted on by the digestive juices.

Feeding of Marine Vertebrates

Broadly speaking fish may be divided into two groups, those which feed on plankton and sometimes on larger swimming animals, and those that feed on bottom-living animals. Of the former, herring, mackerel and tunny are the best known and most important; all swim freely in near-surface waters. The first two feed by straining great quantities of water through the gills where the plankton, especially the small crustacean copepods, is collected in a sieve consisting of

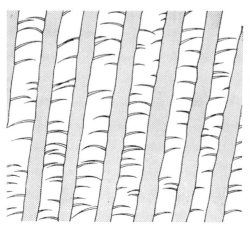

Gill rakers of a mackerel, *Scomber scombrus*, showing the fine spines for sieving off the plankton on which it feeds

parallel series of slender horny projections from the front side of each gill arch known as gill-rakers. They are best developed in the great basking shark which spends the summer in the surface waters feeding on plankton and the winter in deep water growing a new set of rakers for use the following summer. The much more active tunny pursues surface-swimming fish and squids.

The bottom-living fish, such as flat-fish, cod and rays, feed upon the worms and shellfish of various kinds which they seize and crush up with their powerful jaws. It is because the Dogger Bank has an exceptionally rich fauna of bivalve molluscs which are found in great patches sometimes 80 by 32 km, and at a density of between 1000 and 8000 per square metre, that the Bank forms such an ideal region for the growth of plaice. The rays are often able to crush the massive shells of the larger bivalves. One of their number forms the most serious enemy of the pearl oyster on the Ceylon beds although it is no enemy to the pearl diver for this ray is the second host of a tapeworm, the early stages of which live in the pearl oyster and may form the nucleus round which pearls are formed.

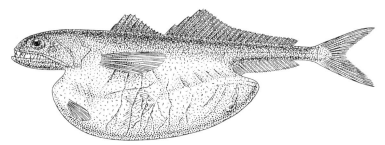

Chiasmodus niger, a voracious deep-sea fish with a fish much larger than itself within its dilated stomach. (×1½)

Other rays do serious damage to oyster beds. More actively swimming fishes such as the cod and haddock also feed on bottom-living invertebrates such as worms and crustaceans. Herring shoals are invariably accompanied by great numbers of dogfish which annually devour large numbers of herring. More unique are the feeding habits of anglers, sluggish fish which live on the sea bottom and have a short, stumpy tail, a comparatively small body and an immense head and broad mouth with inwardly pointed, hinged teeth. Above the mouth is a slender spine, the modified first ray of the dorsal fin, with a tufted end; the angler lies perfectly still with its huge mouth open and daintily waving this tentacle as a lure; thus, far from having to hunt their prey, fish of all kinds, impelled by curiosity or desire for food, come to inspect the lure and are seized in a moment in the open mouth, the recurved teeth of which effectively prevent their escape. Even diving birds have been discovered within anglers. Some of the deep-sea anglers have a highly developed luminous organ at the tip of the tentacle which can perhaps flash on and off. Other deep-water fish have huge mouths and an unlimited capacity for food, their stomachs being capable of such expansion that their owners can swallow animals larger than themselves.

The feeding organs of fishes naturally vary with the type of food. The wrasses, which feed on crustaceans or molluscs, are able to crush them to pieces in their mouths by means of specially powerful teeth in the mouth and throat, but the young of the grey mullet which feed on fine particles and are reported to do considerable damage to the oyster beds owing to the fact that they swallow great numbers of the free-swimming larvæ, have no teeth, really sucking in their fine food. The adults have only weak teeth and browse on encrusting weeds and the small crustaceans in them, the stomach forming a gizzard for crushing up this food. The pipe-fishes also lack teeth and indeed suffer from a kind of permanent lock-jaw; they live on minute crustaceans which they take in by sucking action. The beautiful green or blue parrot-fishes of coral reefs have hard beaks with which they rasp the surface of coral rocks for encrusting sea-weed. The trigger-fishes have jaws and teeth of exceptional strength. They may be used to break open the shells of bivalves.

The feeding of whales forms a strange paradox, the largest of all feeding on plankton, straining it through the frayed fringes of their whalebone plates, while the rather smaller, although immense, sperm whales feed on giant mid-water squid which they seize with peg-like teeth on the lower jaw. Seals and sea-lions feed on fish which they pursue and capture in large numbers, but walruses live on shellfish which they scrape up by means of their large tusks. The sea otter feeds on sea urchins, the large limpet-like abalones and on mussels which it breaks open by banging them against a stone carried to the surface for that purpose.

Suckers and Parasites

The only type of feeders which we have not yet considered are those which live by sucking in fluids or soft tissues. These are not numerous in the sea except as parasites, although a few animals such as some of the little nudibranch molluscs or sea slugs live by sucking in the contents of the cells of green seaweeds. There are a great many parasites which do not dwell within the body of the animal on which they are dependent, but either fix themselves permanently or move about on the outside of the body. These are known as ectoparasites to distinguish them from the endoparasites which live *within* the body of the host. There are many examples of ectoparasites in the marine world, much the commonest being copepod crustaceans which occur on the body of fish and assume all manner of strange shapes; a few of them can be recognized as copepods but in the majority it is only a knowledge of the early stages in the life-history which has enabled us to classify them correctly. Examples of different types of these parasites are given. Some of the less degenerate kinds move about freely on the surface of their host fish, probably living on the mucus and soft skin, but the more degenerate ones bore into the tissues of their host or fix themselves to the gills, or soft skin round the eyes, and have usually piercing and sucking organs for feeding on the blood and soft tissues of the fish.

There are also marine leeches which suck the blood of fish. The most exalted of marine parasites is the hagfish, one of the cyclostomes and a lowly ally of the true fish, which fixes itself by a single tooth into the flesh of its victim and then rasps away the tissues by means of its scaly tongue, sucking in the flesh as it goes.

The most degenerate types of parasites have completely lost their feeding organs; they lie in the body cavity or gut of their host, bathed in nutrient fluid which they absorb directly through the surface of their bodies. There are innumerable examples of which only one—perhaps the most striking—example can be mentioned here. It is quite common to find on the underside of the bodies of various kinds of crabs, a small, round, brownish mass—rather like a tumour but in reality a crustacean parasite called *Sacculina*. It is practically without structure; just a bag containing reproductive products with a branching mass of roots which penetrate the body of the crab in all directions and absorb nourishment from it. Strangely enough this parasite is a lowly relative of the animal in which it lives. It is a crustacean most closely allied to the barnacles, a fact which

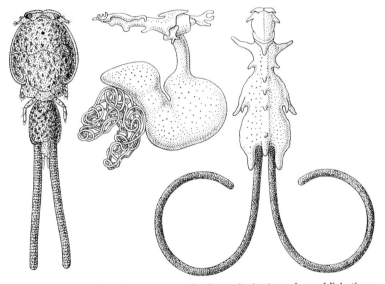

Crustacean copepods which live parasitically on the body surface of fish, those on the right greatly modified. The projecting filaments are egg masses. Left, *Caligus*. Middle, *Lernaea*. Right, *Chondracanthus*. (×4)

has been established by study of its life-history. The eggs, which are discharged freely into the sea from the parent parasite, hatch out into minute pear-shaped creatures, each with three pairs of legs, a single eye, and bearing a little shell exactly like the similar stage in the life-history of a barnacle except that they have no stomach or intestine. These minute nauplius larvæ swim about in the sea for some weeks, when they moult and become cypris larvæ which have a hinged shell almost enclosing the body, a pair of antennae, and some six pairs of swimming legs. For a little time longer they continue to swim about and then must either die or find a crab on which to settle. In the latter case they attach themselves by means of tiny hooks on their antennae to hairs on the body of the crab. The feelers penetrate the base of the hair, after which the legs of the *Sacculina* are cast off and the rest of the body degenerates into a little mass which passes through the hollow feelers into the crab's body. It is carried in the blood stream until it reaches the middle of the body where it attaches itself near the stomach sending out roots in all directions. The main mass then grows larger and larger

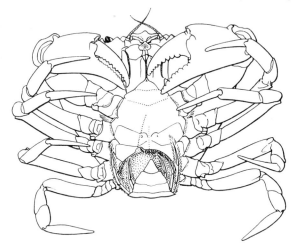

Under surface of shore crab, *Carcinus maenas*, parasitized by the 'barnacle', *Sacculina*, forming a rounded mass below the tail region. (Natural size)

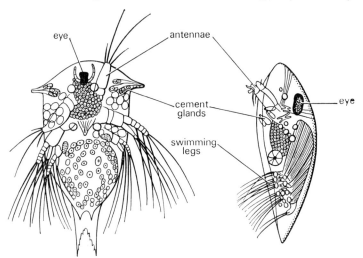

Larval stages of *Sacculina*. Left, free swimming 'nauplius'. Right, later 'cypris' which settles upon and infects a crab. Description in text. (× 35)

until it gets pressed against the inside of the shell about the middle of the underside of the body. The next time the crab moults the parasite pushes its way out before the new shell has had time to harden, and there assumes its final tumour-like shape.

The parasite usually lives some three or four years and from the time it pushes its way to the exterior the host crab never moults; all its reserve of food and strength go to supply the needs of the parasite. It is only when the *Sacculina* dies that the crab is able to resume growth and to moult at the usual intervals. It is a very interesting fact that male crabs which are infected in this way gradually assume the appearance of females, developing a broad instead of a thin abdomen, smaller claws, and certain small legs on the hinder end of the body which are normally only possessed by the females. About 70% of infected male crabs show various stages in this process but infected females on the other hand show no signs of becoming like males. An elaborate theory of the causation of sex has been based on this sex change in parasitized crabs. Details of this are out of place here, but the gist of it is that the presence of the parasite causes the active male to lead the more sedentary life of the female, and to eat more in order to counteract the abnormal drain on its resources. This accumulation of reserve food is thought to affect the animal in such a way as to lead to the appearance of female characteristics even to the production of eggs!

Symbiosis

In a previous chapter we discussed animals which live together, such as gall crabs on corals, and hermit crabs in anemones or with anemones growing on the shells they inhabit. This association has to do not only with defence but with food—thus the anemone in return for the protection it affords the hermit crab may seize fragments of food broken up by the crab, while in the more extreme case of the pea crab which lives within mussels or similar bivalves or within the large sea squirts, the crab intercepts the food collected on the latticed feeding organs of these animals, scooping into its mouth strings of food mixed with mucus. But there is a much more intimate form of association, the significance of which involves provision of food for one or both of the parties concerned. We may have two animals or an animal and a plant living in intimate union the one with the other to their mutual advantage, and often so dependent that one or both cannot exist if separated from the other. This is the type of association known as symbiosis.

The commonest type of symbiosis is the partnership between an animal and minute, green, yellow or brown plants—single-celled algæ—which are taken into the tissues of the animal in large numbers. Many of the simplest animals, the Protozoa, especially the delicate radiolarians, always contain these, as do some sponges, corals and flatworms. The contained algæ are generally known as zooxanthellæ if yellow and zoochlorellæ if green. All are probably capable of living freely but remain in a permanent 'resting stage' when in symbiosis. The advantage of the association to the plant lies not only in protection but in access to supplies of inorganic food represented by the carbon dioxide and the nitrogenous and phosphatic wastes excreted by the animal. The latter has the benefit of this automatic removal of waste products while in some, although not in all, cases the animal obtains food from, or actually consumes, the algæ.

The most striking case of this type of symbiosis is furnished by a small flatworm, *Convoluta*, only a few millimetres long, which is common on the sandy shores of north Brittany and the Channel Islands. It occurs in large colonies which form green patches on the yellow sand, and is an animal of most regular habits, suddenly appearing from beneath the sand immediately after the tide has left it and disappearing just before the tide returns (Plate 27). The green colour is due entirely to the presence of vast numbers of dinoflagellate algae. These are not present in the egg but the animals become infected in a very early stage in development, and if they are kept free from infection by artificial means they fail to develop properly and soon die. Although in early life it is able, like other flatworms, to feed on smaller animals, *Convoluta* soon comes to depend entirely on the starch from the green plants and, as a result of disuse, its digestive organs degenerate so that it becomes impossible for it to feed like a normal animal. This is the explanation of its regular habits for in order to obtain the sunshine without which the plants cannot form starch, it has to expose itself to the full glare of the sun for as long as possible; that is, be on the surface of the sand for the whole time the tide is out; when the tide returns it has to burrow or it would be washed away, the stimulus for descent being provided by vibration from the approaching waves.

But, owing probably to the fact that *Convoluta* needs a more varied diet than that supplied by starch, it begins after a time to feed upon the algæ—to kill the geese which laid the golden eggs—so that they gradually disappear, their owner at this stage presenting the strange appearance of an animal with a green head and a white tail. Finally the little flatworm dies of starvation, though not before it has laid large numbers of eggs. This may be considered a case of symbiosis which has gone too far, for, although the algæ can live freely in the sea and they are by no means dependent upon the flatworms, the latter, after originally no doubt sheltering the algae in return for surplus food, have finally become entirely dependent, essentially parasitic, upon them, and cannot even develop if they are absent.

In the family of the giant clams (Tridacnidæ) the animals farm zooxanthellæ in the greatly thickened and intensely coloured siphonal tissues which project between the open shell valves (Plate 26). In this case the animals retain the power of feeding on plant plankton, like all bivalve molluscs, and, although the contained algæ form a valuable subsidiary source of food, the animal never becomes completely dependent on them, and, in consequence, never suffers premature death like *Convoluta*. In the reef-building corals the zooxanthellæ, as already noted in Chapter 7, are not digested although organic matter certainly passes out of the plant cells into the animal. The precise significance of this, however, still remains to be determined. There is no doubt that the zooxanthellæ act as automatic agents of excretion for the coral and that they also very materially assist in the process of skeleton formation.

Chapter 10

SENSORY PERCEPTION

Animals living in the sea experience an environment very different from that in which we live on dry land. While those animals which live in the intertidal zone of the shore have a periodic existence uncovered by water in which air, temperature and light are much as we experience, those that live below the tidal zone have an altogether different environment. Surrounded always by water, they live at a comparatively constant temperature and under light conditions that vary in intensity and composition with depth, as will be described in Chapter 11.

We can appreciate well enough the range of our own senses, sight, smell, taste, touch and hearing. But what of animals living permanently under water? In recent years much research has been done on this aspect of biology, namely sensory perception.

Let us start with vision. While fish have eyes whose basic structure is very similar to that of our eyes, most of the lower invertebrate animals have very primitive organs for the appreciation of light. These are simple pigmented areas which are sensitive to varying intensities of light but quite incapable of image formation. If symmetrically placed they enable the animal to orientate itself with respect to the direction of the light. By this means they can move from brightly lit areas to those which are dimly lit and thus select their most suitable habitats. Crustaceans, such as crabs and shrimps, have much more complicated eyes built up of a number of similar units each consisting of a lens and retina, like the compound eyes of insects. They are by this means able to gain a mosaic picture of their immediate surroundings which no doubt is converted in the brain to form a useful image. The only invertebrates which have eyes comparable to those of vertebrates, although structurally different from them, are the molluscan cephalopods: squid, cuttlefish and octopus.

But let us begin with the fish. The conditions of light experienced by a deep-water fish living at, say, 1000 to 2000 m will be very different from those experienced by a mackerel or a herring swimming for most of the time in the surface layers of the sea. While these surface-living fish will experience day and night changes rather similar to those on land, the deep-sea fish will be living in perpetual darkness, except for the phosphorescent light given out by other animals.

The main two elements in the eye involved in the perception of light are the lens and the pigment of the retina. These are designed to be best suited for the environment in which we live. Natural selection tends always to favour the organ with the best design for its purpose and it has acted thus on the development of the eyes of deep-water fish. In fish which live in the surface layers of the sea the lens of the eye absorbs ultra-violet rays very strongly. These rays, which can have harmful effects, are quickly absorbed by sea water and can penetrate only a short distance down. Fish which live deeper down are therefore not subjected to them. It is found that the eyes of these deeper-living fish have lenses which are transparent to ultra-violet light. Because they are never subjected to these rays in normal life they do not need the protection given by the lenses of surface-living fish.

The pigments in the retina of the eye can absorb those wavelengths of light,

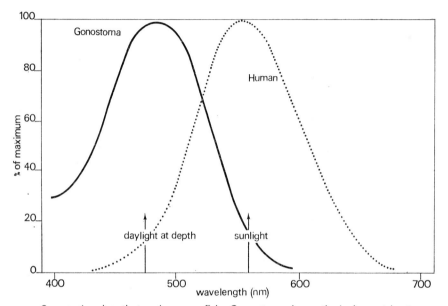

Curves showing that a deep-sea fish, *Gonostoma*, has retinal pigment best suited for the light in deep water whose maximum energy is at about 470 nm wavelength, compared with the human sensitivity curve in sunlight. (By permission of E. J. Denton)

i.e. colours, which pass through the lens and their sensitivity to certain colours is found to depend on the prevailing light conditions. Thus, the absorption of light by the retinal pigments of fishes from different depths has been found to be most efficient for those colours which predominate at these different depths.

Another characteristic of the eyes of some fish is the presence of a reflecting layer made up of crystals of guanin. Like cat's eyes, the eyes of some sharks and rays glow in the dark when a light is shone on them. This is due to the reflecting layer which causes light to pass twice through the retina and thus enables the fish to see better at low light intensities. In some pelagic sharks, such as the dogfish (*Squalus acanthias*) there is a layer of black pigment cells in the eye in which the pigment can move to occlude the reflecting layer. For such fish, which swim into the upper, well-illuminated layers to hunt their prey in the daytime, this is an adaptation to prevent their eyes from shining and becoming conspicuous, for in other respects these fish are well camouflaged by their coloration.

When the eel migrates on its spawning journey from the brightly illuminated waters of a river to the dark waters of the deep sea, it has been found that the composition of the retinal pigment of its eye alters as it leaves the fresh waters so as to fit it for the light conditions it will meet in the sea. Similar, but less marked changes have been found to occur in the retinal pigment of the eye of the salmon when in the smolt stage it starts on its migration from the river in which it was born to the feeding grounds in the sea.

When we consider such fish as eels and salmon we remember the great migrations that they undertake. There has always been much speculation how these fish, and indeed birds also, can find their ways to their correct destinations over these enormous distances. The eel, for instance, has to find its way from different parts of the European coast-line to its breeding area in the Sargasso Sea. Conversely, the salmon makes a return journey from as far afield as Greenland to spawn in the river in which it was born, say in Scotland, Norway or Canada.

It is possible that part of these migrations is passive, the fish collecting in the large ocean currents which will transport them in the direction in which they need to go. But it is now thought that most probably the animal's vision plays a part. They may be instinctively aware of the position of the sun at different times of day and use this information as a compass. Of course, when travelling deep in the water they would be unable to see the sun itself, but the light penetrating into the sea is definitely directional and the brightest area will be that in line with the sun. The light will also be polarized according to the position of the sun and some fish can appreciate polarized light as can turtles and some crustaceans.

Small electric currents are induced in the waters of ocean currents as they pass through the earth's magnetic field. It has been shown that these are within the limits perceptible by the eel and Atlantic salmon, so this may also be a means of navigating.

Until after the second world war it was always believed that the underwater world was silent, and one thought of the dark, silent depths. It is true that certain fish living in shallow waters were known to make noises and that by their sounds the fishermen could locate them. Such fish received appropriate names, such as 'grunters' or 'croakers'. In British waters it was well known that gurnards made a grunting sound when brought on deck in the trawl catch. This sound was made by the air-bladder. The idea of eternal silence was perhaps natural, because man had no means of listening in on the underwater world. But during the war, when underwater sound detectors became generally used in submarine warfare, the listeners were often confused by background crackling noises. So prevalent were these noises at times that they rendered sonic listening valueless. It was soon found that this crackling noise was made by a small crustacean known as the snapping shrimp, *Alphaeus*. It was, in fact, well known to marine biologists that by the action of one of its claws *Alphaeus* could make a sharp clicking sound, and it had consequently been given the popular name of snapping- or pistol-shrimp. But what was not realized was that at times whole populations of this shrimp might make such concerted snapping as to provide a background of crackling noise. Further studies of this behaviour have shown that it is related to territorial habits, the clicks being used to warn off intruders who might want to inhabit the same environment and territory.

As a result of this discovery and with the development of the necessary recording instruments, research was actively directed towards a survey of underwater sound. It was found that beneath the sea surface the environment is anything but quiet and that a multitude of sounds of different pitch and pattern is always present. It is now known that many species of fish make noises, especially at times of breeding. For instance, surprisingly, it was shown that

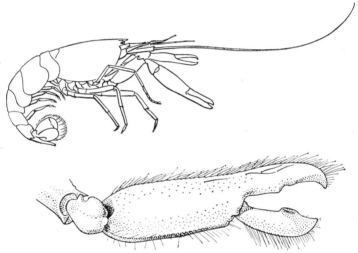

A. The snapping shrimp, *Alphaeus*. B. Enlarged drawing of the claw by which the clicking noise is made

cod make noises at times like grunting. The males grunt during courtship and this stimulates the female to swim upwards to spawn.

Other animals which communicate by sound are whales and dolphins. Often the noises they make are very high pitched, and it has been shown that, like bats in the air, the sounds may be used for echo location. Dolphins can tell the size and even the shapes of different fish by this means.

While whales and dolphins, being land animals that have adopted an aquatic life, have organs for hearing which are well developed, the question arises, how do animals like fish and shrimps hear? A study of this problem has resulted in a much better understanding of the receptor organs in fish and invertebrates.

To start with fish, a very characteristic organ present in all fish is that known as the 'lateral line'. This is an obvious line, often noticeable because of different coloration, which runs along each side of the body of the fish from head to tail. Microscopic examination of this line shows that it is formed by a number of open tubes containing a gelatinous substance communicating with a longitudinal canal whose walls contain sense organs. Each organ has a bundle of minute hairs or cilia which are sensitive to any small movement of the jelly and thus can detect slight differences in pressure due to water movements. As sound is indeed caused by changes in pressure it is evident that the lateral line could thus detect sound, as well as other pressure differences due to currents and so on. Some fish,

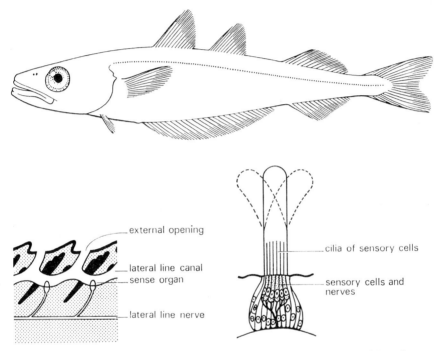

external opening

lateral line canal
sense organ

lateral line nerve

cilia of sensory cells

sensory cells and nerves

The silver whiting, *Micromesistius poutassou,* showing the lateral line (dotted) and enlarged diagrams showing details of a short length of the lateral line in section (left) and a bundle of cilia (right). (Lower diagrams after Goodrich and Fulton)

such as gobies, have a complicated system of similar branching canals running over the head region.

But prawns and other crustaceans have no organs similar to ears for hearing, nor have they lateral lines. On the other hand it is well known that their bodies, especially their legs and claws, are covered in different areas with small hairs and bristles, and that these also occur on their antennae. A careful examination of their disposition and of neighbouring structures of the bristles has shown that they must serve as sense organs for appreciating touch, movement of surrounding water, pressure, and so forth. Thus the snapping shrimp is certainly fitted with organs that can detect the sounds emitted by its neighbours, although as yet the actual organs concerned have not been identified. Since the shrimps show

L

concerted action in their snapping, such organs must exist, because the snapping is so obviously a means of communication.

These sense organs for appreciating pressure changes have, as we have said, bristles or hairs, which give the organs powers of perception. Even very small crustaceans such as copepods (see p. 81) and the arrow worm *Sagitta* are thus equipped. Copepods have numerous minute hairs on their antennae and arrow worms have very small, stiff hairs on their bodies. Such hairs are probably not for hearing, but it is known that they are very sensitive to vibration. It has been shown that they are used to sense the presence of small food organisms when they move near enough. Thus organs which were originally evolved to enable small animals which cannot see to detect vibrations and thus find their food, may well have been the origin of the more highly developed organs in larger animals used for detecting pressure and sound waves.

Animals living in water must have some means of appreciating the quality of the water around them. Land animals can smell and taste; to do this the substance to be detected must first be dissolved in a fluid such as the mucus of the membrane of the nose or the saliva in the mouth. In the sea all substances are already in solution and thus in a state ready for detection by the organs of smell or taste. Such organs are very highly developed in many fish, and they may be sensitive enough to detect substances at molecular levels. Fish will use these organs to test the salinity and other qualities of the water when migrating. Especially will this be so in fish such as eels and salmon which either live in fresh water and migrate to the sea to spawn, or migrate to the sea to feed before returning to the river to spawn. It is now well known that salmon will usually return to the rivers in which they were born and it appears by experiment that they identify the water of the river of their birth by its smell.

Many marine animals use these powers of chemical sensitivity and of touch for quite other purposes than for locating food or sensing the presence of enemies, as already for instance mentioned for the salmon, which can literally taste its home waters. But we have so far been considering animals which can move freely from place to place. A great many members of the animal kingdom in the sea live sessile lives attached firmly to rocks and other solid objects on the sea floor. Some are even attached to other free-living or sessile animals or to seaweeds. All of these animals have, however, as described in Chapter 5, a free-swimming larval existence immediately after hatching. During this period they become members of the drifting community or plankton, in which they may remain for varying periods of time. Eventually they must settle down and their length of stay in the plankton depends very much upon whether they can find a suitable substratum on which they can attach.

Let us take as a typical example the acorn barnacle *Balanus* which encrusts the intertidal rock surfaces. In Chapter 2 the life-history of this animal was described in which there is a 'cypris' stage at which settlement takes place after the free life of the 'nauplius' stage. The cypris having found a suitable solid surface becomes attached and gradually metamorphoses into the adult form. But how does it recognize the right type of material on which to settle and the most suitable spot upon it? It is, for instance, very noticeable that the youngest barnacles appear in fine cracks in the rock and that they tend to be grouped round existing adult barnacles. Recent observations have shown that the antennae of the cypris have sense organs which are both touch and chemical receptors. These young barnacles will select an area of rock on which there are, or have recently been, adult barnacles. Experiments in the laboratory in which the cypris is given a choice of a clean substratum or one soaked in extract from the adult show that the latter is always chosen. In fact a chemical substance has now been isolated from barnacles which evidently has the necessary 'smell' to which the young

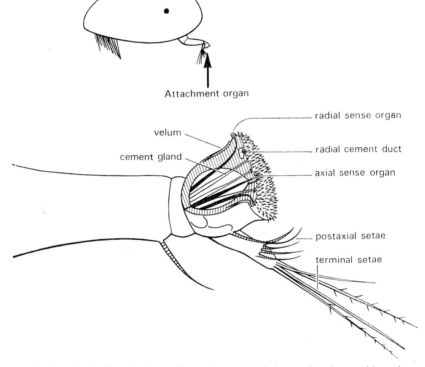

Outline sketch of cypris larva of acorn barnacle, *Balanus*, showing position of attachment organ (above), and enlarged diagram showing details of attachment organ (below). (After Nott and Foster)

barnacle responds. Having thus found a suitable area the young animal must now also select a suitable surface to which it can attach. Feeling round with its antennae it chooses a slight depression or crack in the rock and then immediately brings its cement gland into action. Situated on the antenna, this secretes a substance which hardens almost immediately under water, thus permanently attaching the animal to the rock.

This example is typical of the behaviour of most of the sessile marine animals which in their larval stages seek out suitable situations in the neighbourhood of their own kind. Oysters, for instance, are well known to do this. But many animals are extremely selective, especially those which live commensally in close association with other species. Thus sometimes two species, from different phyla, will associate one with another and with no other species. A good example of this is the anemone which sits on the shell of a hermit crab, and leaves the shell for another when the crab vacates its house for which it has grown too large. Some hermit crabs have another associate. A polychaete worm *Nereis* lives commensally in the shell of a common hermit crab. Again, laboratory experiments have shown the powers of the worm to recognize, and be attracted by, the smell of its host the hermit crab. To test this the worms were placed at the foot of a Y-shaped tube down one branch of which flowed pure sea water, while down the other was a stream of water containing extract from the crab's body. The worm always chose to swim up the latter. Of a similar nature are the methods

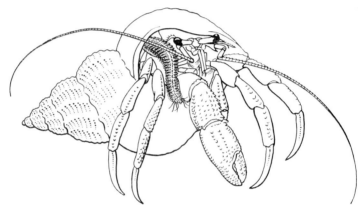

Hermit crab, *Eupagurus bernhardus*, in whelk shell, with the commensal
polychaete worm *Nereis*

which bring the sexes together for mating. The male or the female will give out an attracting chemical, called a pheromone, which guides the opposite sex towards its mate, or else sets off a general spawning reaction.

We have mentioned something about the sense activities of marine animals. Fish, for instance, make noises to communicate one with another. But at times they are silent, and one might perhaps ask, do fish sleep? Although some fish, such as sharks and dogfish, can actually open and close their eyes, bony fish have no eyelids and their eyes remain always uncovered and staring. This has led to the idea that since their eyes are always open they never sleep.

In fact it is now known that most fish at some time, usually at night, adopt attitudes that indicate that they may be sleeping. Some rock-living fish, such as wrasses, are often to be seen lying on their sides. It has now been found that fish in such attitudes can actually be picked up by aqualung divers operating at night. Underwater observations have also shown that pelagic fish like herring, which were thought never to stop swimming, are stationary in the water at

night orientated in any direction and even head downwards. Similar observations have been made of deep-sea pelagic fish. Active and fast-swimming
fish such as the mackerel are now known to congregate and lie on the bottom
in deep water. All these habits suggest that fish, like other higher animals,
need at times to have periods of sleep-like rest.

All these examples that have been given of sensory perception are of animals
having nervous systems, some simple and some highly evolved. All activities

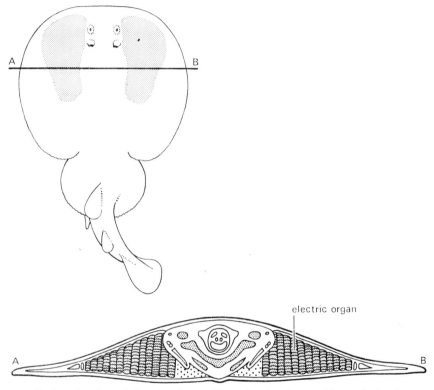

Outline drawing of the electric ray, *Torpedo*, showing positions of electric
organs, and enlarged cross-section of the electric organ at A–B

of the animal resulting from its appreciation of conditions of the environment
are controlled through the nervous system. The receptor organ is connected by
sensory nerve fibres to the central nervous system or brain. This receives mes-
sage from the organ concerned and then, by way of outgoing motor nerves,
communicates with the muscles used in making the necessary movements. The
passage of these messages along nerve fibres is an electrical phenomenon. Thus
all muscle activity is electrical in origin.

Certain fish can store up electricity for use at will. The electric ray *Torpedo*
has a large and highly developed electric organ consisting of many columns of
hexagonal electroplates. These are connected by four large nerves to a lobe of

the brain concerned with these activities. These electric organs are situated on the lateral fins of the fish which, as in skates and rays, are large and wing-like. When the *Torpedo* finds its prey it envelopes it in these great fins and then stuns it by the discharge of an electric current. This is quite considerable and may amount to several amperes at around fifty volts. A number of skates and rays have also been found to possess small electric organs and it is possible that these may be used defensively to drive enemies away.

Let us end this chapter with an account of a remarkable adaptation. In Chapter 8 something has been told of the red colour of prawns in the depths and about luminescent organs. A fish has now been found whose eyes are very sensitive to red light, and whose luminescent light is red. A red prawn in the dark is invisible and if illuminated by blue or green light would still be invisible. But it would reflect red light. This fish preys on red prawns. By switching on its red luminescent organ it reflects red rays from the prawn which becomes immediately visible to its red-sensitive eyes.

Chapter 11

SEA WATER

Chemistry

Sea water is salt to the taste; much more so than fresh water, in which, when we drink it, we are not consciously aware that there is any salt. Yet this is, in reality, present, but in exceedingly minute quantities. For all the oceans derive their saltness from the fresh water which pours down from the land in the form of rivers.

In remote ages one can imagine that the oceans were almost fresh; the saltness of which we are now aware is due to the accumulation of the minute quantities washed down from the land through countless centuries. It has been estimated that there is enough salt in the oceans to yield fourteen and a half times the bulk of the entire continent of Europe above high-water mark.

When sea water is evaporated down until it is dry a white crystalline substance, the sea salt, is left. Over three-quarters of this salt consists of sodium chloride, the remainder being made up of small quantities of bromides, carbonates and sulphates, of sodium, potassium, calcium and magnesium.

Practically everywhere in the oceans of the world the composition of sea salt is to all intents and purposes the same; that is, the proportions to one another of the different components vary very little. But the actual quantity dissolved in the water may vary from time to time and place to place for various reasons. The open ocean water, however, varies in its salinity within very small limits, being almost always between 34 and 36 parts of salt to 1000 parts of water by weight. It is quite natural that in the neighbourhood of land the amount of dissolved salt is lower than in the open ocean, owing to the fresh water which flows off the land diluting it.

In the Baltic Sea, for instance, the salinity of the water is very low, being always below 29 parts per 1000. As we get down to the mouth of the Baltic, however, where it joins the North Sea in the Skagerrak we notice a considerable rise in the salinity due to the mingling of more saline waters coming from the North Sea itself and from water carried round by the drift of the Gulf Stream which penetrates the North Sea round the north of Scotland. Owing to its lessened salinity the stream of Baltic water is quite recognizable as it flows up past the coast of Norway.

On the other hand the salinity of sea water may become considerably higher than that of the open ocean, owing to constant evaporation of the water from the surface and a consequent concentration of the salts left behind. Such conditions are to be found in the Red Sea, where the highest salinities in the world, for open waters, occur, viz. 40 parts per 1000. Here, under the fearful heat of the sun, water is constantly evaporating at the sea surface and there are no rivers flowing down from the land with fresh water to dilute the sea once more. The eastern basin of the Mediterranean is also very salt compared with the open waters of the Atlantic.

(In the case of the Dead Sea, river water has been pouring down for thousands of years into a comparatively small lake in which constant evaporation is taking place. As a result the enormous salinity* of over 200 parts per 1000 has been reached. The difference between this concentration and that of sea water is due to the very small volume of water in the Dead Sea compared with that of the ocean which requires almost incalculable quantities of salt to raise its salinity appreciably.)

In the Atlantic Ocean for the same reasons as those given above, we find the highest salinity in the central part, the Sargasso Sea, and the lowest in polar regions where the lower evaporation and the continual precipitation of rain and snow from the atmosphere combine to dilute the surface layers.

Besides this common sea salt there are also many other substances to be found in solution. In fact there are present traces of almost all known chemical elements. This is only natural when we consider that particles of all kinds of substances must eventually be washed down from land. Amongst these many elements, silver, radium and gold are to be found. The presence of gold has of course attracted men's attention, and certain unscrupulous people have attempted to obtain capital from those who have been misled by the visions of amassing large fortunes out of sea water. Actually, however, gold is present in such minute traces that the cost of its extraction would be greater than the value of the amount obtained. The quantities present have been variously estimated and probably differ slightly from place to place. The highest value obtained is 1 mg of gold in 1 tonne of sea water, or about 1 kg in 3·5 km^3.

It would be even less profitable to attempt to extract the silver, although there is evidently plenty there in bulk; for it has been estimated that there exists dissolved in the oceans 46,700 times as much silver as has been mined all over the world between 1902 and the discovery of America by Columbus, in 1492. This is, in all, 13,500 million tonnes. but it would be a very tedious matter to extract it!

But among the most important substances in sea water are certain 'manurial' salts which are present only in very minute quantities. These are the phosphates and the nitrates, chemicals which we know have a very great nutrient value in the soil on land and are among the chief components of all natural or artificial manures.

We have mentioned in a previous chapter (p. 80) that in the sea, as well as on the land, all life is ultimately dependent on the plants for its supply of food. In the off-shore waters of the sea the plant life is represented almost solely by that great drifting community of plankton plants among which the minute,

---

* Dead Sea salt differs from sea salt in the proportions of its constituents.

unicellular diatoms are the most numerous. It is on the nutrient salts present in solution in the sea that these diatoms depend for their food. On land the plant extracts phosphates and nitrates from the soil by means of its roots. The fine 'root hairs' absorb the nourishing salts from the minute interstices among the soil particles where the moisture holds them in solution. In the case of diatoms, however, there is no need of any specialized structure such as the root to take up these valuable substances, for the whole body is bathed in the sea water from which the food is absorbed. More will be said in the next chapter on this subject and the effects of the presence or absence of the manurial salts on the seasonal cycle of life in the sea.

Besides these 'solid' salts in the sea there are also present certain gases dissolved in the water. Animals cannot live without oxygen. We extract the oxygen that we breathe in with the air into our lungs, and it is carried round to supply the various parts of our bodies. But in the sea we never see fish coming to the surface to take in a gulp of air as the marine mammals, such as whales, do. They have no need to do so, for oxygen is everywhere present, dissolved in the surrounding water, and fish have special structures, the gills, for extracting it. All the larger marine animals have gills or some such specialized structure for this purpose. But in the case of the very small animals, in which the area of the body surface is very large compared with the actual volume of the body itself, oxygen can be absorbed anywhere over the surface of the body.

There is probably no part in the open sea where oxygen is not present in solution in sufficient quantity to support life, although there are deep layers in which the oxygen content is low. Isolated cases are known, however, where there is no oxygen. Such conditions are to be found in the deeper waters of the Black Sea. Here we have a layer of light surface water down to about 180 m, below which is heavy water. The upper layers are of low salinity, owing to the fresh water brought down by the Danube and other rivers, but the deeper waters have a high salinity; hence the difference in weight of the waters. Thus there is at a certain depth a layer where the light surface water and the deep heavy water meet. This forms a kind of boundary layer between the two, and the waters above and below this depth do not mix one with another. Therefore, although the oxygen supply in the upper layers may be very high, there is no means by which this gas can be transported into the deeper layers. As a result there is no oxygen below about 180 metres, the only gas present being the stinking sulphuretted hydrogen. No living animals are therefore present below this depth. The only organisms that can live are certain bacteria that do not require oxygen, and it is these bacteria that are responsible for the production of the sulphuretted hydrogen. Such a condition is very rare in salt water. Even in the very greatest depths in the open ocean life is still possible. For the absence of oxygen is a result of stagnant conditions, and in the open ocean such can hardly ever be the case, owing to the continual circulation of water in the great ocean currents.

While we breathe in oxygen we send out into the air again the product resulting from its utilization, carbon dioxide. All animals are therefore constantly giving this gas out to the surrounding water and it is everywhere present. At the same time it is used up again in the upper layers by the floating plants which build up sugars from it under the action of sunlight, restoring oxygen once more to the water.

Physical Properties

Water being the medium in which all marine animals live, there are certain of its physical properties which influence the animals themselves or tend to modify the environment under which they live. On land, for instance, the temperature of the air or the barometric pressure may induce profound changes in our bodies.

The surface temperature of the sea changes markedly from place to place. In the tropics it is hot compared with the polar regions. The highest temperature recorded in the sea is 35·5°C (96°F) in the Persian Gulf and the lowest 2·2°C (28°F) in polar regions. Between these two limits all temperatures are to be found. Albeit this range is small compared with the great temperature differences which occur on land, the highest being 68·8°C (136°F) in Tripoli and the lowest −77°C (−127°F) in Antarctica. But although the changes are not as great as on land, they are of marked significance in the lives of the animals living in the sea. For whereas most land vertebrates are warm-blooded and have special means for keeping the temperatures of their bodies constant, in the sea most animals are cold-blooded and must take up the temperature of the surrounding water. If our temperature rises only a few degrees the chemical reactions are gone through at a dangerously increased rate and give rise to fever. We can easily imagine, therefore, that the passage of an animal into water two or three degrees warmer than that in which it had been living may have a profound effect upon it. It is probably for this reason that the boundaries to the distribution of many animals in the sea are those of temperature (page 62).

Water requires a tremendous amount of heat energy to raise its temperature; the amount of heat necessary to raise the temperature of a cubic metre of water by one degree (C) would raise by the same amount 3000 cubic metres of air. This is why in the summer the sea never has time to reach the high temperature of the surrounding air except sometimes just at the surface.

For the same reason water is very slow to give up its heat when once it is gained, and we notice that in winter the sea is far warmer than the air, and acts as a reservoir of heat. This explains why the climates of oceanic islands and coastal lands are much more equable than those of countries in the interior of great continents. The influence of oceanic currents upon climate is well shown by a comparison of that of the British Isles, whose shores are bathed by a branch of the Gulf Stream, with the climate of Labrador which lies in the same latitude.

At the same time any heat received at the sea surface is imperceptibly slowly carried downwards into deeper layers, and in general there is a decrease in temperature with depth. Usually the effects of the sun's warmth are not felt to a greater depth than 550 m. Actually the heat rays are rapidly absorbed by the upper few centimetres of water and any warmth that is transported to deeper layers is mostly brought about by the mixing of the warmer with the colder water.

Owing to this very slow warming up of the sea there is a lag in the seasonal change of temperature, and while in temperate latitudes the hottest time of the year is June, the water does not reach its maximum temperature until August. Now we have seen that the deeper layers only receive their heat by mixing with the warm upper water and while there is a lag between the raising of the temperature of the air and that of the water there is a further lag in the warming of the deeper water, until at about 90 m we get a complete reversal of seasons. There the hottest part of the year is in December, that is mid-winter, and the coldest about May or June. Below 200 m there is no seasonal change whatever in temperature, and, year in year out, the conditions are uniform. From this depth downwards the temperature gradually falls until in the ocean abysses it remains always at somewhere near the freezing point.

The actual weight or specific gravity of the water depends on its temperature and upon its salinity, the warm water being lighter than the cool, and the fresh water lighter than salt. In the tropics, therefore, the high temperature makes the water light, but at the same time, owing to evaporation, the salinity is high which tends to make the water heavy.

The pressure in the sea varies with the depth. At every ten metres depth the pressure is increased by one atmosphere. In the great depths of the ocean the pressures are therefore enormous, as much as a thousand atmospheres. Yet there is no part of the sea in which animals cannot live. At first sight it seems remarkable that any living creature could endure such enormous pressures, but we must realize that that is their natural environment. We, on land, live always with a pressure of 6·4 kg to 6·5 cm² on our bodies, but we are not conscious of it; when the barometer rises by 2·5 cm the increase of pressure on our body surface may be as much as one tonne, yet this is not noticeable.

The great pressures in the depths of the ocean have a slight effect on the density of the water, but because water is almost incompressible the increase in density with depth is extremely small. It can in no way be sufficient to support a persistent and erroneous popular belief that, owing to the increasing density, objects sinking will find their own level before they reach the bottom, a level in which the density of the water is the same as theirs and below which they cannot sink because the density of the water becomes greater. Sir Wyville Thomson in *The Depths of the Sea* remarks, 'There was a curious popular notion, in which I well remember sharing when a boy, that, in going down, the sea water became gradually under the pressure heavier and heavier, and that all the loose things in the sea floated at different levels, according to their specific weight: skeletons of men, anchors, and shot and cannon, and last of all the broad gold pieces wrecked in the loss of many a galleon on the Spanish Main; the whole forming a kind of "false bottom" to the ocean, beneath which there lay all the depth of clear still water, which was heavier than molten gold.'

It has been suggested for instance that the *Titanic* thus sank to a false bottom; but under the increasing pressure sealed air chambers would become 'imploded' or smashed inwards, so that eventually there would be no air to buoy the ship up and it would be simply a mass of solid iron and wood, the combined density of which would far exceed that of the sea water surrounding, so that the ship would be bound to sink to the bottom. Sir John Murray said, 'During the *Challenger* expedition, after a funeral at sea, the bluejackets sent a deputation aft to ask if "Bill" would go right to the bottom when committed to the deep with a shot attached to his feet, or would he "find his level" and there float about for evermore ?'

It is said that a man at 3600 m would bear on his body a weight equivalent to that of twenty locomotive engines, each with a long goods train loaded with pig iron. This was said however in 1873 when engines and trains were considerably smaller than nowadays. One of the effects of living at these great pressures is that many animals when brought up quickly in the trawl break to pieces owing to the sudden reduction of pressure.

Most sea animals must be able to stand a considerably wide range of pressures as they are able to make large journeys up and down. Whales when harpooned have been known to 'sound' as deep as 800 m. Many small animals living in the drifting community make upward journeys of 50 to 100 m or more almost every night of their lives, retiring to the deep levels again in the daytime. In the laboratory small unicellular animals have been subjected to pressures of as much as 600 atmospheres, without suffering any apparent harm.

One of the most important conditions for life is light. Without light there would be no plant life, for it is by the help of the sun's rays that the plant can build up the necessary starch and sugar from the carbon dioxide in the air or in the water. Practically all animals are ultimately dependent upon plant life for their food, and if in the sea the rays of light from the sun were cut off immediately and completely by the surface water, no plants could survive there and it is certain that our oceans would not contain that wealth of animal life in which they all abound.

The amount of light to be found at any depth in the sea depends upon the altitude and strength of the sun, on the weather conditions, and upon the amount of sand and sediment present in the water, that is to say its turbidity. Much light is reflected from the sea surface especially if there are waves, because the light rays glance off the sloping sides of the waves. It is often almost impossible to look towards the setting sun over the sea because of the dazzling path of reflected light, and we must realize that this light is not being reflected only along the path between us and the sun, but all over the sea surface, because from whatever place we look towards the sun we shall always find this path of light. The lower the sun is in the heavens the more its rays are reflected. Thus we see that it is seldom that all the light from the sun penetrates the actual sea surface; such a thing only happens when the sun is vertically overhead and the sea itself is as calm as glass. But the rays of light that do pass through the surface cannot penetrate right to the bottom in very deep water. Out in the great oceans the darkness on the sea floor, many thousands of metres beneath the surface, is absolute. This is because the light itself is absorbed by the water. But all the different colours of which white light is composed are not absorbed to an equal extent. The red rays for instance are absorbed very quickly indeed. It is a matter of several metres only before all the red light has gone. It is the blue and violet light that penetrates farthest and in clear ocean waters William Beebe noticed, on one of his bathysphere dives, that every trace of visible light had vanished at 610 m (2000 ft) (see p. 5). Some years ago experiments made in the Sargasso Sea showed that if a photographic plate was exposed for eighty minutes at a depth of 1000 m (3280 ft) it was blackened by the light rays; but a plate exposed for 120 minutes at 1701 m (5578 ft) was not affected.

An excellent example of the absorption of light by the sea water is furnished by the famous Blue Grotto of Capri within which everything is enveloped in the purest blue light because the only light that can enter has to pass first through the water which practically fills its narrow entrance. In its passage through the water much of the red light and some of the yellow and green is absorbed, and the only light that can come once more above the surface of the water to illuminate the interior of the cave is composed to a very large extent of blue rays. The blue colour of the sea also owes its origin to this phenomenon, for the colour of the water is due to the reflection of light upwards from the small particles suspended in the water itself. The light reaching a particle at a given depth is thus reflected upwards and has to pass once more through the depth of water it has already traversed in its downward journey. Much of the red and yellow light will become absorbed on this upward journey if this has not already happened on its downward passage, and it is mostly blue and green light which can survive to appear above the surface once more and give the sea its typical colour. As all these rays of light are being absorbed in their downward passage it is natural that the actual strength of the light is gradually diminishing the deeper it goes. In the open ocean the strength of light is already too weak at a depth of 200 m to support much plant life (see p. 5), and below this depth few living plankton plants are to be found. Nevertheless this upper layer of water, 200 m in thickness, is sufficient to support the tremendous wealth of plant life that forms the pasturage of the sea, and it is on the rain of dead plants and organisms that have fed on them that the animals in the dark ocean depths largely depend for food. It is hard for us to realize what the actual strength of the light is at different depths in the sea; but we might give as an example that in the open waters of the English Channel the light at a depth of about 18 m corresponds to that found in the heart of an English beech wood, and this is quite dim. At such a depth in the Mediterranean, however, the light would be very much stronger, not only because the sun itself is so much more powerful but because the water is clear and lacks that sediment of sand and mud that so discolours the water in the English Channel.

Currents

The great oceans of the world are not merely masses of stationary water. There exists within them a system of gigantic currents whereby the water is continually circulating and moving from place to place.

Amongst these currents is that known as the 'Gulf Stream', a name familiar to all on account of its bearing on the climate of north-west Europe. In order to understand roughly how the Gulf Stream takes its being and how it moves, it is necessary to consider the whole of the North Atlantic Ocean. The Gulf Stream is a great oceanic current that receives its name from its place of origin,

the Gulf of Mexico, whence it issues through the Straits of Florida as a stupendous river of warm blue water 80 km in width and 640 m deep.

The forces that set and keep this great mass of water in motion are varied,
and among these the polar ice, the sun's heat, the trade winds and the rotation of
the earth all play their part.

This current is really only a portion of a system of currents in continuous
circulation in the Atlantic Ocean, a system which cannot be said strictly to
begin or end anywhere. We will however confine ourselves here to the North
Atlantic.

In the region just north of the equator the surface waters are warmed by the
fierce heat of the tropical sun and their salinity is raised by constant evaporation.
Upon these warm saline waters are continually blowing those persistent north-
easterly winds known as the 'North-east Trades'. Their action aids a natural
tendency of the surface waters to move in a westerly direction towards the north
coast of the South American continent as the North Equatorial Drift, which
flows on into the Caribbean Sea and Gulf of Mexico. Here the waters become
piled up so that the surface level is raised by as much as 18 cm above that of the
ocean water outside the islands of the West Indies in the Sargasso Sea. Under
such conditions the water must flow somewhere in order to maintain its equilibrium. Its only place of egress is through the Florida Straits and here it issues
as the mighty Gulf Stream, the actual current through the straits being sometimes known as the Florida Stream. From here the Gulf Stream runs in a
northerly direction along the coast of America at a speed of about four knots,
giving off eddies on its way (Plate 24).

Owing to the rotation of the earth there is a tendency for any moving particle
to be deflected to the right in the northern hemisphere. This action is felt by the
Gulf Stream and it moves northwards with an increasing trend towards the
right or east, so that by the time it has reached latitude forty degrees north it is
flowing due east across the Atlantic. At this point it has lost a certain amount of
its speed and has widened out considerably. At the same time, coming into less
heated climes, it has cooled down to a certain extent.

Close to the Labrador coast, however, a potent force comes into play. Here
are vast masses of drifting ice floating down from polar seas on the cold Labrador
currents. These ice floes meet the warmer water and cool it. Cooled water is
heavier than warm water and must therefore sink, and as it sinks it is replaced
by more warm water near the surface. This process continues while the ice is
melting, so that it, so to speak, attracts the warm waters to it, asking for destruction. The mass of all the ice in this region thus exerts a considerable power
and deflects a part of the Gulf Stream which splits off from the east-going
current and moves up north over the Newfoundland Bank and on to the Norwegian Sea.

The remaining east-going stream continues on across the Atlantic; part,
owing to the earth's rotation, still bears to the right until eventually it returns
to its place of origin in equatorial waters, thus completing a circle. The remainder, sucked north to replace water moving south from polar regions and aided
by prevalent south westerly winds, moves on towards the British coasts as the

North Atlantic Drift. Still bearing to the right it bathes the coasts of Ireland and then the western coasts of Scotland. Passing the north of Scotland it meets again that portion which had been deflected near the Labrador coasts and the two carry on together to the north of Norway, a small branch turning south round the Shetland Islands into the North Sea.

On reaching Arctic regions this originally warm surface water has been cooled down nearly to freezing point. It has however still a larger salt content than the water over which it has been flowing, and is therefore heavier at the same temperature. It therefore sinks, and then creeps slowly back towards the equator as a deep bottom current which rises once more just north of the equator to replace the water that is being driven westwards on the surface by the trade winds.

Such in simple outline are the main features of the oceanic circulation in the North Atlantic. In the South Atlantic a slightly different circulation exists on

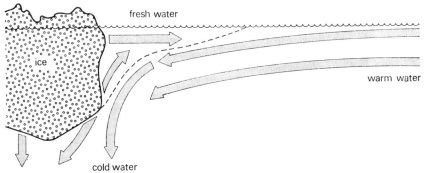

fresh water

ice

warm water

cold water

Diagram showing the movements of the surrounding water caused by melting ice

account of the configuration of the land masses; and the two systems become linked up in the region of the South and North Equatorial Drifts.

In the Pacific, too, there is a system of oceanic currents much after the manner of that in the Atlantic, the Japan Current or 'Kuroshio' (Black Stream), corresponding to our Gulf Stream.

Whenever a mass of water is moved another mass must take its place. There are thus many deep currents and countercurrents in the oceans not mentioned in the above simple outline.

In these great circulations there are to be distinguished two types of water movement, the 'drifts', and the 'currents' or 'streams'.

The drifts are brought about by wind action, and are horizontal movements of the surface waters more or less in the same direction as the wind is blowing. Currents on the other hand are translocations of bodies of water caused by a definite head of water. In the Gulf of Mexico such a head of water is produced in the manner mentioned on page 164, with a result that a current flows out through the Straits of Florida, the Gulf 'Stream'. On reaching more northern latitudes, however, the driving force on the Gulf Stream ceases to be a difference in water level, but it is now sucked in towards the ice in polar regions and considerably aided by persistent south-west winds which drift the surface water along in a north-easterly direction, so that the waters that bathe British coasts are more correctly termed those of the North Atlantic 'Drift' than the Gulf Stream.

These types of water movement, drifts and currents, are to be found always around north European coasts. There is for instance a prevalent drift of surface water up the English Channel and through the Straits of Dover into the North Sea. In fact 'drift bottles' (i.e. special bottles used for measuring water movements [see p. 182]) liberated in the western end of the English Channel have been known to travel in the surface drift on to the Swedish and Norwegian coasts in just over a hundred days, a distance of about 1127 km. One indeed was picked up on the north shore of Norway after having travelled a distance of 2317 km at a speed of about 12·2 km a day. In these instances, however, there was no actual movement of water mass at anything approaching this speed.

When a strong wind has been blowing for some days on to the shore, the surface water then becomes piled up on the shore on which the wind is blowing,

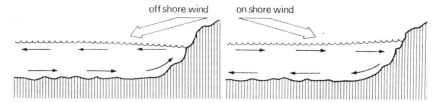

Diagram to show the currents set up by off- and on-shore winds

with the result that a head of water is produced and water flows outwards in the deeper layers to replace the water that has been driven in by the wind.

There remains one other type of water movement, and this is movement of bodies of water in a vertical direction, either upwelling or sinking. Sinking water masses are produced in various ways, such as in the case of water of a uniform salinity cooling in its surface layers, when the density of this surface water increases and it sinks downwards. Upwelling of water occurs when a deep current meets a submerged bank or the shelving bottom of coastal regions; it can also be brought about by persistent off-shore winds, which blow the surface water outwards and this is replaced by water from below. The upwelling of water can be of the greatest importance in some areas, for the deep waters of the sea abound in those nutrient salts, the phosphates and nitrates, and when water from deeper levels rises to the surface it brings with it this store of nutriment to replenish the impoverished surface waters (see p. 175). This is why life is so abundant in the region of submerged shelves and in coastal areas, and for instance accounts for the great fisheries off Peru.

Tides

Apart from these great circulatory currents in the ocean there are other periodic movements brought about by the attraction of the sun and moon on the waters. These are the tides.

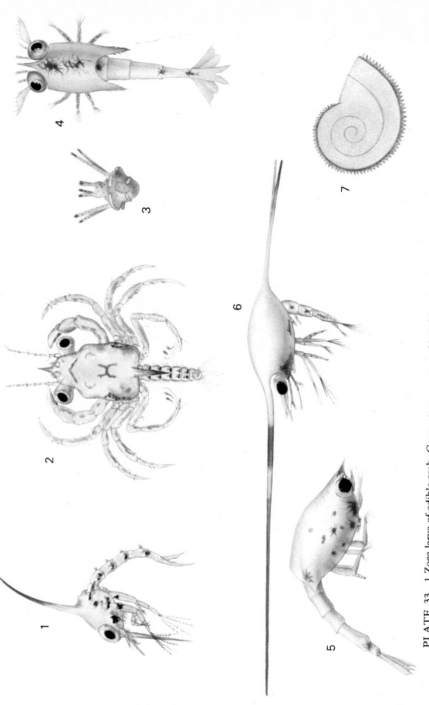

PLATE 33   1 Zoea larva of edible crab, *Cancer pagurus*, × ca. 20.   2 Megalopa larva of edible crab, *Cancer pagurus*, × 20. (After M. V. Lebour)

Larval stages in plankton   3 Larva of sea urchin, *Echinus esculentus*, × 20.   4 Larva of squat-lobster, *Galathea*, × 18.   5 Larva of hermit crab, *Eupagurus*, × 18.   6 Larva of porcelain crab, *Porcellana*, × 18.   7 Shell of larval mollusc, *Echinospira*, × 18. (After F. S. Russell)

(above) Deep-sea trawl (Photo: Institute of Oceanographic Sciences) (right) Mid-water net (Photo: Institute of Oceanographic Sciences)

PLATE 34

Grab for taking samples of sea bottom, (below left) Open. (below right) Closed

PLATE 35

(above) Hauling trawl in North Sea (Photo: F. S. Russell)

(above) Stern trawler (Photo: Brooke Marine Ltd, Lowestoft)

(*above*) Date-mussels, *Lithophaga lithophaga*, boring into calcareous rock, ×⅔ (From *Die Mytiliden*, Fauna und Flora des Golfes von Neapel, No. 27. By kind

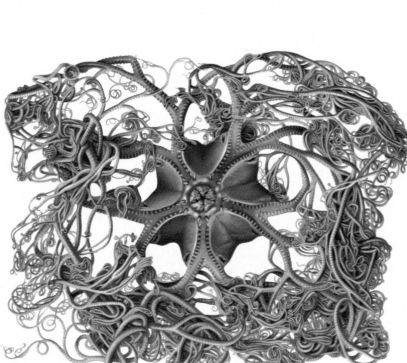

(*above*) Basket star, *Gorgonocephalus agassizi*, showing much branched arms of this complex brittlestar, ×½ (Reproduced by courtesy of the Institute of Oceano-

PLATE 36

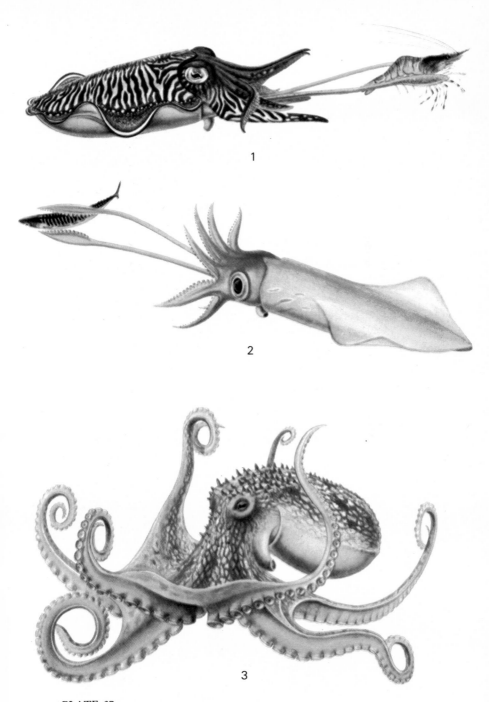

PLATE 37

1 Cuttlefish, *Sepia officinalis*, ×½. 2 Squid, *Loligo forbesi*, ×¼. 3 Lesser octopus, *Eledone cirrhosa*, ×½ (Painted from life by Peter Stebbing)

PLATE 38

(below) Catch on board. The fish have just dropped out of the cod-end

(above) Trawl alongside. Notice gallows and otter-board in background, glass balls for floating head line, and cod-end floating with mass of fish

(below) Mullet net, Alexandria, Egypt (Photo: F. S. Russell)

PLATE 39

Eggs of herring,
× ca. ½ (Photo:
Ministry of
Agriculture,
Fisheries and Food)

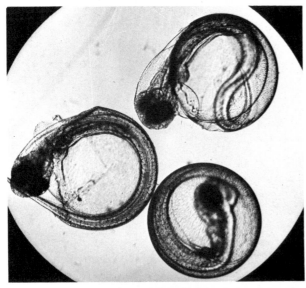

Plaice hatching
from eggs, × ca. 15
(Photo: Ministry of
Agriculture,
Fisheries and Food)

Herring eggs
enlarged, × ca. 9
(Photo: Ministry of
Agriculture,
Fisheries and Food)

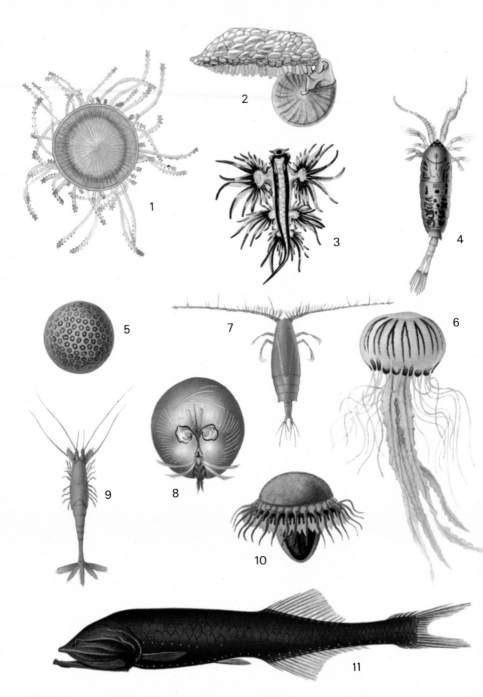

PLATE 40 Showing colours of pelagic animals from different depths from surface downward. 1 *Porpita mediterranea.* 2 *Janthina fragilis.* 3 *Glaucus atlanticus.* 4 *Anomalocera patersoni.* 5 *Halosphaera viridis.* 6 *Chrysaora.* 7 *Gaetanus kruppi.* 8 *Gigantocypris agassizi.* 9 *Eucopia australis.* 10 *Atolla* sp. 11 *Cyclothone livida* (Modified after A. Steuer's *Planktonkunde* by Peter Stebbing)

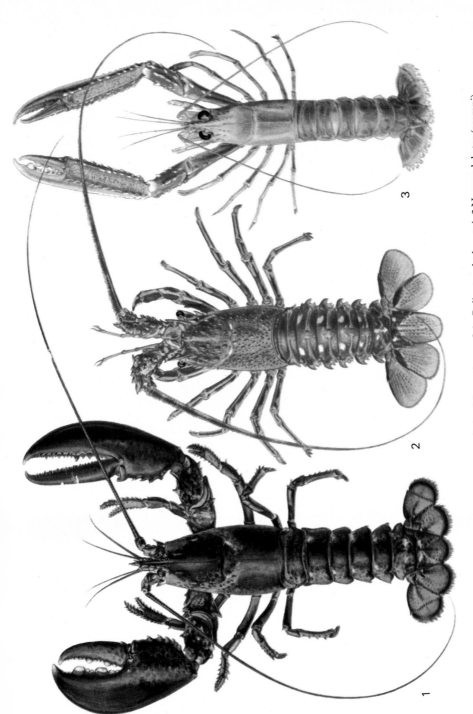

PLATE 41   1 Lobster, *Homarus vulgaris*, ×½. 2 Spiny lobster or crawfish, *Palinurus elephas*, ×½. 3 Norway lobster (scampi), *Nephrops norvegicus*, ×½ (Painted from life by Peter Stebbing)

(above left) Otolith of plaice showing growth zones, × ca. 20 (Photo: Ministry of Agriculture, Fisheries and Food)

(above right) Scale of herring in its fourth winter, × ca. 8 (Photo: Ministry of Agriculture, Fisheries and Food)

PLATE 42

(centre right) Young plaice caught in one hour from pool shown below, × ca. $\frac{4}{5}$

(below right) Pool in Gannel Estuary, Newquay, Cornwall (Photo: F. S. Russell)

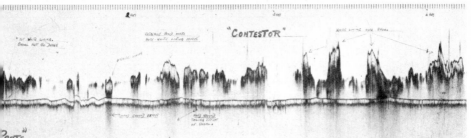

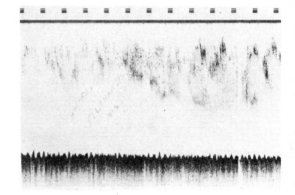

(*above*) Echo trace of sprat shoals (Photo: Kelvin Hughes)

(*right*) Echo trace of pilchards and mackerel (Photo: Kelvin Hughes)

PLATE 43

Vickers deep-sea submersible 'Pisces' (Photo: Vickers Ltd)

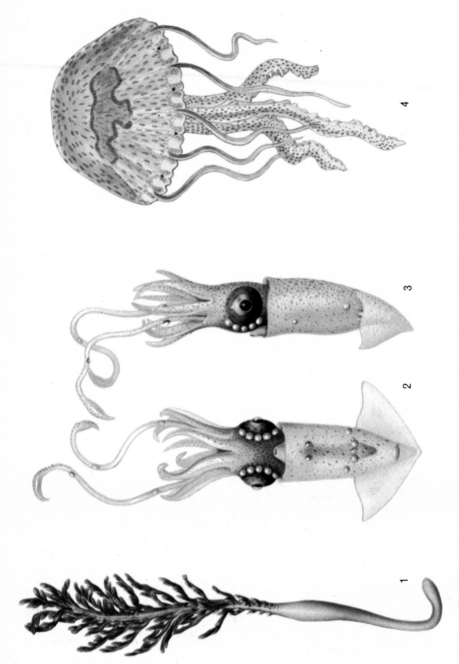

PLATE 44   Luminescent animals. 1 Sea pen, *Pennatula phosphorea*, ×¾. 2 Underside of deep-sea squid, *Thaumatolampus diadema* showing arrangement of light-producing organs or photophores, ×1¼. 3 Side view. 4 Jellyfish, *Pelagia noctiluca*, natural size. (Sea pen reproduced by kind permission of the Carnegie Institution of Washington; the other two creatures

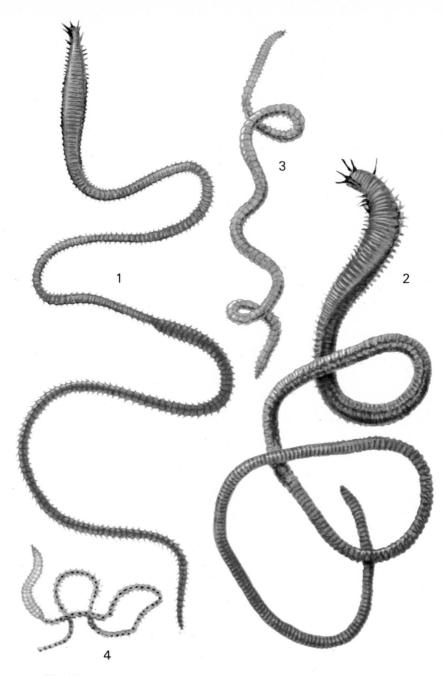

PLATE 45 ATLANTIC PALOLO WORM, *Eunice fucata*

1 Mature male, hinder portion ready to break off. 2 Immature male, ×2. 3 Female sexual portion broken off. 4 Empty female sexual portion (By permission of Carnegie Institution of Washington)

Oyster park at Arcachon, south of Bordeaux, exposed at low tide with collectors in the foreground (Photo: H. A. Cole)

PLATE 46

Collecting oysters from exposed surface of parks off the Ile d'Oléron south of Rochfort, Biscay coast of France (Photo: C. M. Yonge)

Collecting mussels from the buchots on the muddy bottom of the Anse de l'Aiguillon, north of La Rochelle (Photo: C. M. Yonge)

Mussel pontoons in the Ria Vigo, north-west Spain (Photo: C. M. Yonge)

Hawksbill turtle, *Eretmochelys imbricata* (Photo: Zoological Society of London)

PLATE 47

Green (edible) turtle, *Chelonia mydas* (Photo: Zoological Society of London)

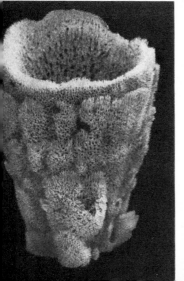 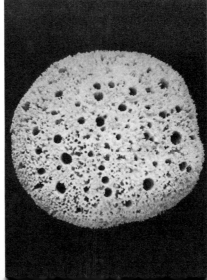

(left) Turkey-cup sponge. *(right)* Honeycomb sponge

Japanese oyster culture. *Crassostrea gigas* growing on stakes in shallow water, shown exposed at low tide (Photo: C. M. Yonge)

PLATE 48

Suspended oyster culture in Japan. Surface view of rafts from which growing oysters attached to scallop shells are suspended on wires (Photo: C. M. Yonge)

The mass of the moon exerts an attracting force on the particles on the earth. This force is greater on those particles which lie closest to the moon, but it is

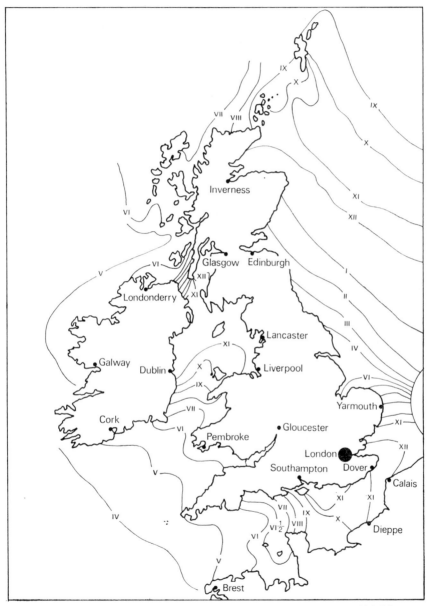

Map showing the progress of the tidal wave round the shores of the British Isles. The Roman numerals indicate the times of high water as the wave moves along the coast

extremely small, being only one ten-millionth of the earth's pull on those same particles. Let us imagine the earth to be stationary and covered uniformly all over with water. The pull of the moon on the water immediately beneath it is so small as to have no effect in drawing it vertically away from the earth's surface. But the farther we go from this point on the hemisphere directed towards the moon the greater does the moon's horizontal pull become. It does not take nearly so great a force to make water slide horizontally over the earth as to draw it vertically upward, and the moon's attraction is sufficient to do the former. The result is that water is drawn over the surface of the earth from all around and towards the point beneath the moon so that the water becomes piled up there. Where this bulge of water occurs the tide is high.

On the hemisphere of the earth pointing away from the moon exactly similar forces are acting to cause high tide at the point farthest from the moon. How these forces come about can be proved mathematically, but it is somewhat beyond the scope of this book to enter into an explanation.

In this way there are two high tides simultaneously on the earth, one at a point beneath the moon, and the other on the other face of the earth opposite to the first. These two high tides cause the water lying between them to be drawn away, so that at two points lying on either side of the earth midway between the two high-tide regions there will be low tides.

But the earth is revolving, so that approximately once each day the moon exerts its influence on every meridian on the earth's surface in turn. Under these conditions the points at which high tide occurs change with the changing position of the moon, and a 'tidal wave' sweeps round the surface of the globe. (This must not be confused with the popular and wrongly named 'tidal wave' of great dimensions and destructive force, which is usually a wave caused by a submarine earthquake and quite unconnected with the moon's attraction.) If the earth were covered uniformly with water down to a depth of 85 km the speed of this wave around its surface would be about 1610 km an hour. Owing to the comparative shallowness of our oceans, however, it has only about half this speed.

The surface of our earth is not uniformly covered with water, but possesses great land masses which jut out into the ocean and tend to complicate the results of these tidal forces so that a truly satisfactory theory of the tides has not yet been evolved. It is, however, safe to say that the primary cause is that given above.

The only region on the earth where the ocean forms a continuous band around the globe, uninterrupted by land masses, is in the southern ocean and there is a theory that around this belt sweeps the great tidal wave, a 'primary wave'. This wave is supposed to give off secondary, or 'derived waves', which move in a northerly direction up the other oceans. Such a wave moves up the Atlantic Ocean at a speed of about 805 km an hour. In the open ocean the wave is only 0·6 to 0·9 m in height, so that the tides in oceanic islands are very small. But as the water shallows round the edge of the ocean the speed of the tidal wave is enormously reduced by friction and the height of the tides becomes greater. Thus the height of the tide at the Azores is about 1·5 m, that around the Scilly Isles 5 m, and that at Swansea 8 m. As the tide runs up narrow channels the water tends to be bottled up and therefore the highest tides are found in such channels as the mouth of the Severn, where the tide may reach nearly as much as 15 m at Chepstow.

There are produced around British coasts two high tides in the twenty-four hours and two low tides, but owing to the fact that the moon is also moving round the earth the interval between one high tide and the next is not exactly twelve hours, but on the average twelve hours and twenty minutes. Thus the time of high water is not the same every day, but there is a progressive change, the high

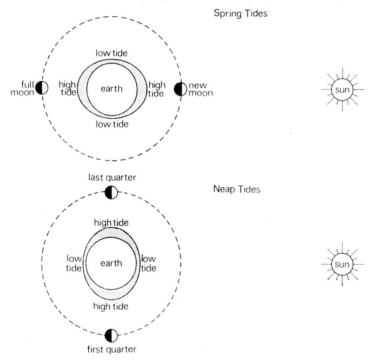

Diagram illustrating the action of sun and moon in causing spring and neap tides

tide of tomorrow afternoon being about forty minutes later than that of this afternoon. So throughout the years there is practically no time of day at which high or low water may not occur at any place.

Together with this change in the time of the tides there is also to be noticed an alteration in the range of the tides, the distance between high- and low-water marks becoming progressively greater throughout a certain period and then correspondingly smaller. These periods are known as the periods of 'spring' and of 'neap' tides.

During the 'spring' tides (Saxon: *sprungen*, 'to bulge') the low-water mark is lowest and the high-water mark highest; that is, the range between the tide-marks is greatest. During the neaps the reverse is the case, the tide coming in and going out only a short distance. Besides the attraction of the moon there is also a smaller attraction exerted by the sun. It is because of the sun's attraction that the spring and neap tides are caused. When the sun and moon are both exerting their pulls in the same direction the force will be at its greatest and the big tides or springs will occur. This happens shortly after both new and full moon. But the range of the tides will be greatest at new moon, because then the sun and moon are both on the same side of the earth; and the spring tides are less at full moon, the sun and moon being then opposite one another. When the forces of the sun and moon are at right angles to one another, i.e. at the periods of half moon, the range of the tides will be least and neap tides will result.

The tides must not be confused with 'tidal streams' which are the horizontal currents moving the water which forms the progressing tidal wave. In the open oceans the wave, as already mentioned, is only 30 cm or so high, so that the tidal stream is imperceptible; but near the coasts where the tidal wave increases in height the tidal stream may be considerable.

Waves

Waves are an integral part of the sea. It is rarely, if ever, the case that the sea is so calm that there is not the slightest ripple, or failing that a long low swell, the aftermath of a heavy storm or the herald of wind to come. Waves are of great importance in helping to keep the surface waters of the sea mixed, while in the tidal zone they have left their mark in the many adaptations shown by littoral animals for protection against the pounding surf.

Waves are generally the result of the action of wind on the sea surface, from the faintest ripple caused by the light airs of summer to the tumultuous mountains of water raised by the full force of a winter's gale (Plate 30). The size of the wave depends upon the strength of the wind and upon the distance through which the wind can act. The largest waves, therefore, occur in the oceans where there has been sufficient distance over which the wind has blown. Mr. Vaughan Cornish in his book on Ocean Waves states that a fetch of some 1000 to 1500 km (600 to 900 miles) is required to produce the largest waves. The height of a wave is the vertical distance between the summit of the crest and the deepest part of the trough. Figures for the height of waves in relation to the speed of the wind are given by Cornish as follows: With a wind of 50 km per hour the height of the wave is 6·4 m, at 80 km/h it is 10·7 m, at 102 km/h, 13·7 m, and above 121 km/h about 21 m; these figures are under conditions when the wind has had a long fetch. Waves of these dimensions are now recorded regularly by special measuring instruments, and examples are 21 m south of Iceland and 15 m on three days in a fortnight in the western approaches to the English Channel. It is, however, possible that waves up to 30 m in height may occur when a number of waves coincide in their passage.

The waves on the sea at any one time are the result of an integration of a number of wave patterns due to local winds and to storms at varying distances away. Methods are available whereby these patterns can be recorded and analysed. In calm weather swell waves have been recorded on the Cornish coast which must have come 9660 km from storms near the Falkland Islands in the

Southern Ocean. It had probably taken these waves about ten days to travel this distance.

The wave in the open ocean is of the type known as an oscillatory wave. That is, while an undulation passes through the water, after the wave has passed, the water particles are still where they were and have not received any movement in a horizontal direction. Actually the movement of the water particles is a circular one, backward and upward on the lower half of the wave front, then forward and upward to the summit of the crest, forward and downward for the upper half of the back of the wave, and then backward and down to the hollow of the trough (see figure). When these oscillatory waves approach the shallow water of the coast their shape changes. They become shorter and higher and the advancing front of the wave steepens and the water particles now have a forward motion. Eventually the increasing speed of the crest causes it to topple over in

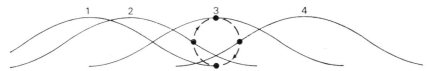

Diagram illustrating the movement of a particle on the sea surface caused by the passage of a wave. 1, 2, 3, 4 are the successive positions of the wave crest and of a particle on the sea surface at the same time

a mass of seething foam to form a breaker and add to the grandeur of the surf. When the wind is off-shore water is blown off the top of the breaking crest in the form of spray.

There may be waves present, but no wind. Such waves are known as 'swell', and they are waves that have travelled beyond the windswept area or have been left on the face of the troubled ocean after a storm has passed (Plate 30). A combination of swell and wind-waves can often be seen.

There are, besides, the so-called 'tidal waves', the products of land or submarine earthquakes and meteorological conditions. These may be waves of very large dimensions produced by an earthquake on the sea floor; or, when an earthquake occurs in coastal regions, the sea may recede far beyond the normal low-tide mark and then hurl itself with relentless fury back on to the land, bringing death and destruction with it.

Such waves are known as 'tsunamis' (Japanese *tsu*, harbour and *nami*, wave). Recent examples are the landslide produced by an earthquake in Alaska in 1958, which caused waves to sweep to a height of 530 m on the headland opposite; and the disaster in the Bay of Bengal in 1970 when conditions due to the monsoon caused a large body of water to be driven by wind into a narrowing bay. In the open ocean earthquake waves may travel at 480 to 640 km per hour.

Ice

In the polar regions the surface of the sea is frozen. When the surface water freezes it does not usually attain a greater thickness than 2 m in a year. When first formed it is known as 'floe ice'. Cracks in the ice, widening into lanes, cause the separation of the ice into floes which drift at the mercy of wind and current. Under certain conditions these floes become driven together and under the great pressure they become packed together to form a mass with uneven surface, floes becoming tilted up on end and forced half out of the water, the whole presenting an appearance of jumble and disorder. Such ice is known as 'pack ice'.

In the open ocean, often far from polar regions, 'icebergs' are met with. These are in no way connected with the floe or pack ice, but are derived from glaciers. The majority of those met with in the North Atlantic come from the great glaciers of Greenland. These solid rivers of ice flow slowly into the sea, where partly by wave action, and partly by uneven adjustments of weight, great masses break off to form the icebergs. The actual process of breaking off is known as the 'calving of an iceberg'.

These castles of ice, often fantastic in shape, float down into the Atlantic on the Labrador Current, a menace to passing shipping. Only a small portion of the mass of ice is visible above water, as it floats with almost eight-ninths submerged.

Recently a new method has been tried for destroying these floating perils. A boring is made into the side of the berg and a charge of 'thermite' inserted, which, on ignition, gives off an intense heat. After several hours the berg breaks up owing to the uneven stresses and strains set up by the temperature changes occurring in its interior after the explosion, the cracking of the berg into pieces sounding like a series of gun reports.

Since the tragic sinking of the *Titanic* by an iceberg, an Ice Patrol Service is on constant duty to give warning to passing ships of the presence of the bergs and to study their drift and movements.

Curiously enough one of the first signs of an approaching iceberg is a rise in the temperature of the water. The surrounding sea water on bathing the ice becomes cooled and hence heavier, and sinks, more warm water flowing in to take its place. But the fresh water from the melting of the exposed part of the berg pours down off its sides into the sea. Owing to its freshness it is considerably lighter than the sea water and floats at the surface. Similarly water from the ice melting below the sea surface rises. Fresh water heats up more readily than salt water, so that this lake of fresh water which surrounds the iceberg on all sides gains more heat from the sun than the sea water.

Chapter 12

OCEAN SEASONS

One of the most obvious phenomena on land in temperate climes is the rotation of the seasons. If we were suddenly let loose from a prison in which for many years we had lost count of time, and if our course were then directed to a country garden, a wooded valley or a mountain marsh, we could read at a glance by the presence of flowers, birds or insects the season of the year. We can follow this endless daily change in our gardens, in the woods, and in ponds, but to many the seasons of the sea are a closed book, save for the changes to be seen in that narrow fringe around our shores, the tidal zone.

It is for this reason that the subject is given a chapter to itself, to emphasize the fact that there are seasons in the ocean, and that in the causes of the seasonal changes lie some of the most fascinating problems of marine research.

Perhaps nowhere can the signs of the seasons be more clearly read than in that drifting community of plant and animal life known as the plankton. Anyone who takes collections with the tow-net at regular intervals of time in coastal waters in temperate regions will be struck at once by the unexpected suddenness with which the composition of the catches may change from week to week. He will see an ever-changing panorama of life.

In the early months of the year, March and April, when the sun is climbing in the heavens and its light increasing in strength, we note a very great change in the plankton from that of the previous months. Under the influence of the benevolent rays of the sun, the tiny floating plants, the diatoms, thrive and multiply exceedingly. Each single-celled plant divides in half within the space of a day; the two cells thus formed reproduce in turn to give rise to four, so that in the course of a week or a fortnight their numbers have risen prodigiously. If we draw our tow-net through the waters of the sea in spring we find that its meshes become clogged with these little diatoms (Plate 22), and, although singly almost invisible to the naked eye, they now tinge the surface of the net green, so countless are their numbers. Then indeed is the pasturage of the sea most rich.

Just about this time too there is a veritable outburst of animal life in the plankton. In the early months of the year a great number of all kinds of animals start to breed. As we have already mentioned, there are few animals around our shores whose early days are not spent drifting freely in the water layers. The presence of these temporary members of the plankton becomes very noticeable in April, when a tow-net catch will be found to consist almost exclusively of these babies of the sea. There will be larval stages of starfish, molluscs, worms and crustaceans (Plates 30, 33). Close around the rocks are amazing swarms of the minute free-swimming young of the acorn barnacle which covers the surface of the rocks between tide-marks (Plate 30). Now is their time to enjoy a free life, for within a month they will have fastened themselves to the rocks by their heads, there to remain kicking food into their mouths with their feathery legs until they die. Farther from the shore, stretching over kilometre upon kilometre of water are the young of that ubiquitous plankton copepod, *Calanus*, and mingled with them are the young of countless other closely allied but smaller species. Young fish, too, now become abundant. All these are born in time to partake of the rich pasturage drifting in the sea. The most minute of the animals will feed directly upon the diatoms, and in their turn will fall a prey to their own larger enemies.

By May or June the great crop of drifting plants is considerably diminished; most of it has been eaten, but a quantity may also have sunk to the sea floor to form food for the large population of animals living on and in the upper layers of the sea bed. This plant community, or phytoplankton as it is called, gives place in the summer months to a community of animals, the zooplankton, forms of which are now growing up and feeding one upon another. Throughout the summer the appearance of this population is always changing, new forms being set free into the water by their parents and others disappearing to take up their abode on the sea bottom.

So the succession of life passes on until in the autumn, about October, there is another somewhat surprising outburst of plant life, yet never so great as that of the spring. Soon after this the plankton rapidly dies down, until in the winter it is at its poorest, only a small number of animals surviving to tide over the lean months and give rise once more to their numberless progeny in the following spring when the sea wakes up from its winter sleep.

One of the most interesting problems in marine research has been the search for the causes of this succession of life in the sea, and it was only in the first half of this century that the underlying principle was brought to light. It is evident that on the quantity of plant life present depends the number of animals that can exist, and that therefore the great changes in wealth of animal life can be traced to the increase or decrease of plants. The countless animals which reach maturity during the course of the summer months do so at the expense of the great flowering of drifting plants that occurred in the spring. These for some reason die down in the summer, to be followed by a smaller and shorter outburst in the autumn. To explain these seasonal changes we must go back still further, and find the causes for these periodic maxima of plant abundance.

It is common knowledge that to obtain good crops on land we must manure the ground; of first importance in manure for plant growth are those inorganic substances, the phosphates and nitrates. Sea water contains these in minute quantities in solution. At the beginning of March, and indeed throughout the previous winter months, these salts are present in greatest quantities, but it is not until the spring that the diatoms are able to utilize these nutrients to any great extent; for plants are dependent for their activities upon light, and during the winter the sun's rays are very feeble; owing to the sun's low altitude little of its light penetrates into the water and much is reflected from the surface. In the spring months, however, the sun mounts in the sky and its light increases in strength. The sun's energy now becomes available for the plants' activities and they start to grow. With the large available fund of manurial salts the growth is rapid; but the food supply is not inexhaustible and within a month it is almost all used up and the great crop perforce dies down. Throughout the summer months there may be very small sporadic outbursts of plant life where the water has perhaps been temporarily enriched in nutrient material by the presence of large shoals of animals whose excreta enrich the surrounding water, but the renewed vigour of the diatoms in the autumn still remains to be explained. The cause of this has been found to lie in the physical properties of the sea water. During the hot summer months the surface water became warmed and we have seen on page 160 that owing to the bad conductivity of the water this heat is only very slowly dissipated into the deeper layers. The result is that a layer is formed at the surface with a temperate two or three degrees (C) higher than that of the water beneath. Between these two water masses there is a very narrow layer in which the temperature changes very abruptly from that of the upper water mass to that of the lower. This is known as the 'discontinuity layer' or 'thermocline' and owing to certain physical properties of water it becomes extremely difficult for the upper and lower water masses to mix one with another, thus forming an almost impenetrable boundary, which can only be broken down under the action of very severe gales. The depth at which the layer is formed becomes deeper as the summer advances, but never gets much below twenty metres in temperate waters.

As a result of the formation of this discontinuity layer we find the clue to the poor plant life during the summer giving rise to a fresh outburst in the autumn. For while all the nutrient salts have become used up in the surface layers a fresh supply is being formed in the deeper layers from the sinking down of dead and dying animals and from other organic remains. But owing to the inability of the upper and lower layers to mix, this nutrient supply is mostly cut off from those plants living in the surface waters above the discontinuity layer where the light conditions are best for growth. During the summer months, however, a certain amount of the nutrient-rich deeper water passes through the discontinuity layer into the surface waters by turbulence and allows growth of very small flagellates on which the zooplankton feeds. In the autumn, when the surface layers begin to cool down, the waters become once more of the same temperature from top to bottom. All the water masses can now be mixed freely and this is soon brought about by the autumnal gales. As a result fresh supplies are brought up into the surface layers where there is still sufficient light for the plants to grow actively and a renewed flowering of the diatoms takes place. Soon, however, with the onset of winter, the sun's light becomes too weak and the plants die down until they are wakened once more into life in the following spring.

From the changes in the amount of phosphates in the waters of the English Channel, off Plymouth, a minimum estimate has been made of the actual extent of this diatom crop during the year. Assuming a depth of 68 m the minimum annual yield of diatoms would be as much as 5·6 t (wet weight) in the water layers lying beneath 0·4 ha of the surface. It is interesting to compare this with the quantity of crops obtained on land. For instance, between the years 1895 and 1904 the average annual yield of potatoes was 4·9 t per acre (0·4 ha) the highest

value for any one year being 5·9 t. Over the same period that for turnips and
swedes was 13·4 t.

These seasonal changes in the quantity of plankton are reflected in the growth

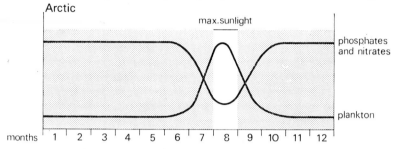

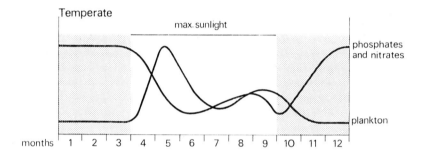

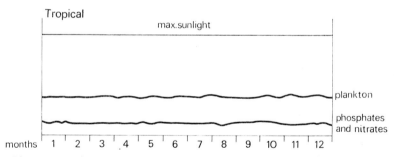

Diagram showing the seasonal sequence in the abundance of plant plankton
in relation to the quantities of nutrient salts in arctic, temperate and tropical
latitudes

of all the larger animals in the sea. A very great number of the small invertebrates living on the sea floor might be termed annuals. Spending their early life
in the drifting plankton community in the spring, they soon settle to the bottom
and reach maturity in the course of the summer. The countless young that never
survive to maturity have nourished other growing animals, which in turn have

fallen prey to the larger creatures. It is probably largely for this reason that we find that most of the fish in temperate waters make the greater part of their growth during the summer. Plankton is then present in greatest quantity for those pelagic fish such as the herring and the mackerel which feed directly on it. Then, too, at its greatest is the toll of life among the young molluscs and crustaceans consumed by such bottom-living fish as the plaice. It is natural that during the period of greatest feeding most growth takes place and this is shown very distinctly on the scales of many fish. A reliable index to the age of these fishes is afforded by the number of wide and narrow zones shown on their scales, the wide zones corresponding to the period of greatest growth which generally takes place in the summer, and the narrow zones to the period of poor growth in the winter.

Such then is the cycle of life in the sea in temperate regions. In polar waters, owing to the short period in the year when there is sufficient light for plant growth, there is only one main outburst of life which occurs during the summer months. In tropical waters, on the other hand, there are not the marked temperature changes that occur in temperate seas, nor does the strength of the sun's light penetrating the sea surface vary to such an extent during the course of the year. Conditions are such that at all periods of the year planktonic plants can grow actively and the marked seasons of polar and temperate regions are absent.

Thus, in general, there is a progression from north to south of a single outburst in polar waters, through a double outburst in temperate seas, to a more or less uniform production throughout the year in tropical regions. In many parts of the world, however, there are dry and rainy periods, monsoons, and other weather conditions that alternate with unfailing regularity during the year and these secular changes have their effect upon the life in the sea. Monsoons especially are effective in causing upwelling along the coasts of nutrient-rich water.

As conditions throughout the year in tropical waters are generally favourable for the growth of plankton plants the supply of nutrient salts may be kept at a low level. Since, however, their reproduction goes on at a faster rate in warm water and it is continuous throughout the year, the total production during the year may be as great as in higher latitudes. But at any one time the standing crop of plankton in the tropics will appear much smaller than in spring and summer in high latitudes.

There are also in tropical, as well as temperate, waters animals which, if not giving an index of the season of the year, show without fail, by their spawning habits, the times of the lunar months. It is quite a common occurrence to find that marine animals will breed at certain fixed stages of the moon. In Egyptian waters, for instance, there is a sea urchin that breeds during the nights when the moon is full. But perhaps one of the most surprising cases of this kind is supplied by a small marine worm from the tropical seas near Samoa in the Pacific Ocean, the palolo worm (Eunice viridis). This worm can be regarded as a veritable sea calendar. All the year round it lives in holes and crevices among rocks and coral growth on the sea bottom. But true to the very day, each year the worms come to the surface of the sea in vast swarms for their wedding dance.

This occurs at dawn just for two days in each of the months, October and November, the day before, and the day on which the moon is in its last quarter; the worms are most numerous on the second day, when the surface of the ocean appears covered with them. Actually it is not the whole worm that joins in the spawning swarm. The hinder portion of the worm becomes specially modified to carry the sexual products. On the morning of the great day each worm creeps backwards out of its burrow, and when the modified half is fully protruded it breaks off and wriggles to the surface, while the head end of the worm shrinks back into its hole. The worms are several centimetres in length, the males being light brown and ochre in colour and the females greyish indigo and green. At the time of spawning the sea becomes discoloured all around by the countless floating eggs.

The natives are always ready for the spawning swarms as they relish the worms as food. They catch them by dipping them up in special baskets and so greatly do they esteem them that the native chiefs send them as presents to those living inland. The worms are eaten either cooked and wrapped up in bread-fruit leaves, or quite un-dressed. When cooked they are said to resemble spinach, and taste and smell not unlike fresh fish's roe.

From remote antiquity it has evidently been the custom of the natives to watch for them, for the natives of Fiji, who call them mbalolo, have incorporated them in their calendars. They call the parts of the year corresponding to the October and November swarming periods mbalolo lailai (little) and mbalolo levu (large), the November swarms being the greater of the two.

The natives first note the approach of the season by the appearance of the scarlet flowers of the aloalo (*Erythrina indica*). They watch for the flowering of other plants until the seasea (*Eugenia*) is in bloom. Then they look for the moon being just on the horizon at the dawn of day, and on the tenth morning the palolo worms appear. However, sometimes the extra lunar month throws out their calculations.

In Savaii the approach of the palolo is heralded three days beforehand by the appearance of the malio or land crabs (*Gecarcinus*) which march down from the mountains to the sea in swarms.

The natives of the Gilbert Isles say that the palolo is a production of the coral, growing out of it, and they call it nmatamata or the glistener; it appears in that locality in June and July.

There is also a closely allied form, the Atlantic palolo (*Eunice fucata*) (Plate 45), whose habits are very similar. The swarms of this worm have been studied in the Gulf of Mexico and they always appeared within three days of the time of the moon's last quarter between 29 June and 28 July. The sexual portions begin to rise to the surface at least two hours before sunrise; by sunrise they are present in countless numbers and when the first rays of the sun strike the water they break up and discharge the eggs. The dying bodies sink to the bottom where they are eagerly devoured by waiting fish and by three hours after sunrise none is to be seen.

At Amboina in the Malay Archipelago there is a similar worm, the wawo, which swarms on the second and third nights after the full moon of March and April.

The Japanese can also boast a palolo; they call it bachi (*Ceratocephale osawai*). In this case it is the front end of the worm which becomes modified and breaks off. The swarming occurs during the nights immediately after new and full moon in October and November and the fishermen catch them by means of lights to which the worms are attracted; they use them then as bait.

It is hard to understand what can be the cause of these curious phenomena. It has been thought that the stimulus for spawning might be found in the state of the tides. Experiments have shown, however, that probably the tide can have no influence, for worms placed in floating tanks spawn naturally at the usual times, and in this case the worms could have had no means of telling the state of tide either by the pressure of the water or the speed of its movement.

It seems more probable that the stimulating influence lies in the presence of the moon, though quite how it acts is hard to imagine. More, as yet, we cannot say but we hope that in the future an explanation may be forthcoming. At any rate these lunar rhythms, probably of very common occurrence among marine animals, are receiving much attention from those interested in the science of the sea.

Chapter 13

METHODS OF OCEANOGRAPHICAL RESEARCH

Until the nineteenth century the great oceans of the world were comparatively little known to man. That is to say, although by then much of the actual geography of the seas had been mapped out, the world that lived beneath the surface of the water and the conditions to be found there were as a closed book to mankind.

At this day our knowledge is considerably increased and a study of the ways and means by which the discoveries have been made discloses the obvious dependence of oceanography upon the advance of other branches of science and the improvements of mechanical engineering and electronic instrumentation. For exploring the greatest depths of the ocean, for instance, the advantages to be gained by the employment of steam or motor winches for hauling in the great weights of gear and rope are manifest. The introduction of wire cable to replace the hemp ropes in 1874 inaugurated a noteworthy advance. The ropes required for heavy work had, of necessity, to be extremely thick, and the space needed to stow many thousands of metres of such material was great, whereas very much thinner wire provides the same strength and takes up very much less room, and can be kept wound upon the drum of the hauling winch.

One of the first investigations in a study of the sea was to find the depths at which the sea floor lay below the surface from place to place. While originally the purposes of sounding were to aid in navigation, and the depths only in comparatively shallow waters were sought, with the introduction of submarine telegraphy a study of the contours of the bottom at all depths became necessary for the laying of the great transoceanic cables.

In shallow water the usual method is to heave overboard a lead weight attached to a length of rope which is marked off at intervals with pieces of leather and other materials to indicate the depth in metres. The lead weight is usually from 4 to 6 kg and has its bottom hollowed out to form a cup into which is put some tallow. On striking the bottom the bits of sand or pebbles adhere to this grease and the navigator is thus enabled to discover the nature of the bottom over which he is passing.

For sounding in deep water a machine was always used and the sounding line itself was made of thin steel cable, which passed over a grooved wheel, each revolution of which corresponded to a known length of wire. In very great depths it becomes impossible to feel when the lead has struck the bottom, because the weight of the many metres of wire in use is sufficient to continue unwinding the drum on which it is coiled. But by the employment of a brake, however, which can be tightened up to counteract the increasing weight of the wire as it runs off the drum, the machine becomes so delicate that the moment the weight touches the bottom the brake acts and the depth can be read off a dial. When a very large amount of wire is off the drum, its own weight comes dangerously near its breaking strain and there is a risk that, when it and the lead weight attached are hauled in, a break may occur. This difficulty was overcome previously by employing a specially designed weight which consisted of a number of separate sinkers, and on the lead striking bottom these were released and the weight to be hauled up was thus considerably reduced. To save time nowadays the wire is usually sacrificed.

At the present time a much improved method of sounding has come into practice. This method is an outstanding example of the application of knowledge gained by research in other branches of science, being a combination of the

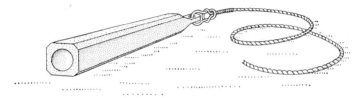

Sounding lead, showing hollow for 'arming' with tallow

use of sound and electricity. It is known as 'echo sounding' and the principle of it is that a sound is sent vertically downwards to the bottom and a very delicate instrument picks up the sound once more as it is reflected back from the bottom; in other words it receives the echo. Now the rate of travel of the waves of sound through water is known (about 1494 m per second) and by noting the interval of time between the first transmission of the sound and the reception of its echo it is possible to calculate the distance through which it has passed and hence the depth of the bottom. In shallow water this interval of time becomes extremely small, and it is only by the use of elaborate electrical apparatus that the time can be measured. With such an instrument a ship can travel swiftly through the water, recording continuously as she goes.

The original word to 'sound' (which is considered to be a form of the Old English word 'sund', meaning 'swimming') would nowadays appear to have been very happily chosen and could not express better what is actually being done when a depth measurement is being taken.

Of equal importance in navigation comes the necessity of knowing the speed and direction of the ocean currents. The bearing of currents on the distribution of the life drifting in the water layers is a problem that has received considerable attention in modern fishery research. Information can be gained about currents both directly by noting the movements of floating and drifting objects or actually measuring the speed and direction by special instruments, and indirectly by a study of the physical and chemical condition of the water itself from place to place.

In olden days much of the knowledge gained about ocean currents was acquired by the mariners themselves noting the position of drifting wreckage or derelict ships and as reports came in from different vessels it was possible to mark off on a chart the route taken by a drifting wreck and therefore to gain some idea of the general trend of the current. There was, however, a danger in basing too definite a conclusion on such observations, as the wreckage always had a portion above the water level exposed to the winds so that the path taken by such a wreck was therefore caused by a combination of wind and current. For many years information has been gained by using specially constructed bottles which are thrown overboard at specified positions. They are known as 'drift bottles' and they are used to study the water movements at the surface or near the sea bottom. The surface bottles are so weighted that they float almost completely submerged, and thus the least possible area is exposed to wind action. The bottom bottles are made just the slightest bit heavier than water and have fixed to them about 45 cm of stiff wire; this wire trails along the bottom and holds the bottle clear of stones and other obstructions that might impede its movement as it is carried by the current (Plate 31). In each bottle is placed an addressed postcard with directions printed on it in five different languages instructing the finder to place it in the nearest letter-box after having filled in the place and date of recovery. Some of these bottles travel for enormous distances; for instance one which was liberated in the middle of the English Channel was picked up away on the coast of Norway, having journeyed a distance of 2317 km in 190 days (see p. 166). Some may also remain in the sea for a very long time. One was returned to the Plymouth laboratory on 1 October 1971 which had been picked up on the Dutch island of Terschelling. This was one of 390 bottom drifters put into the North Sea farther south on 17 November 1904! Of these, 272 had however been trawled up again by 31 August 1906.

Plastic now replaces glass for making special drifters for studying currents (Plate 31).

Currents can also be measured directly by means of special recording instruments. Several instruments have been invented, but there is one in general use, the principle of which is as follows. The machine is lowered to a given depth in the water and to it is fixed a vane which at once sets the instrument facing in the direction of the current. The current then acts on a small propeller which it rotates. Fixed to the instrument underneath is a small circular tray divided into a number of compartments corresponding to the divisions on a compass. After every so many revolutions of the propeller a little lead shot is released which falls on to the centre of a grooved magnetic needle, runs down it, and falls off its north point into the compartment of the tray that lies immediately beneath it, according to the direction of the current. After a certain time the instrument is hauled up on board, and by counting the number of shot in the different compartments of the tray the number of revolutions of the propeller can at once be calculated, and from this the speed of the current; also at the same time, by noting the number of shot in each separate compartment, the direction in which the current has been flowing is at once given. Electrical current meters are also now used.

A recent innovation for direct measurement of deep-water currents in the open ocean is a specially designed apparatus which can be made to float at any desired depth. This carries a transmitter which sends sound waves back to the ship and by very accurate methods of navigation its drift with the current can be followed and timed.

The indirect methods of attack depend upon the fact that the physical and chemical characters of the water vary from place to place. For instance the Gulf Stream water is notably saline, and by an examination of the salinity its course can be roughly traced as it moves northwards. The physical state of the water and its chemical composition have an extremely important bearing on the life contained therein, and the temperature, density and saltness of the water are being kept continually under observation by those engaged in marine biological research in different regions.

The chief instrument for studying the temperature of the sea is the thermometer. The study is a fairly simple matter when it is only necessary to find the temperature of the water at the sea surface. Water is merely dipped up in a wooden bucket and an accurate thermometer placed in it for two or three minutes; it is important to use a wooden bucket rather than a metal one, because metal is such a good conductor of heat that in a short time it may materially affect the temperature of the water contained in the bucket and so lead to errors.

But it is quite another matter when we wish to know the temperature at 100, 200 or even 2000 m. The water down to a certain depth becomes cooler as one goes deeper, and the first difficulty to be overcome therefore is that if a thermometer is lowered, say to 100 m, it will pass through warmer and warmer water as it is hauled to the surface, and by the time it is examined it will be registering a different temperature from that actually occurring at 100 m. This difficulty is overcome by using specially designed bottles that are thoroughly insulated. In addition, by means of a weight known as a 'messenger' which can be sent down the wire, the bottle, which is lowered open at both ends, can be closed (Plate 31). On its downward journey the water merely flows through it, so that when it is closed it takes a sample of water from that depth at which closing took place. Owing to the insulation the water sample thus obtained keeps its original temperature, and this is then read off from a thermometer fixed in the bottle itself. This is a reliable method when a knowledge of the temperature is required only in shallow waters, but a further complication steps in if one wishes to study the deepest water layers.

The water at this depth is under very great pressure, but when the sample has reached the surface the pressure is very much reduced. Now if water is compressed its temperature is slightly raised, and conversely if the pressure is reduced the water becomes cooled. In the passage of a sample in the insulated water bottle from a great depth to the surface, the water, cold as it was at the start, will be slightly colder when the observer reads the temperature. This error must therefore be counteracted. Accordingly a 'reversing thermometer' is used,

N

that is one in which there is an S-shaped bend, so that if it be suddenly turned upside down the thread of mercury is broken and a permanent record of the temperature is obtained. The thermometer on reaching the required depth is reversed by means of a weight which slides down the wire and releases a spring catch.

By attaching the thermometer to a reversing water bottle the insulating water bottle is no longer necessary and has been superseded. Incidentally, the depth from which the sample has been taken can also be determined by comparing the readings of a thermometer protected by thick glass from compression at great depths with those of one not so protected.

To save speed in measuring the temperature an instrument known as the bathythermograph is used. By using the principle of the expansion and contraction of metal at different temperatures combined with a pressure gauge a graph of temperature with depth is scratched on a small smoked glass slide. Thus, a continuous record of the temperature is obtained as the instrument is sunk in the water and drawn up again.

With the insulated water bottle samples of water are obtained which, besides giving information on the temperature, can be drawn off into clean, stoppered bottles to await future chemical analysis. The water is subjected to extremely delicate examinations which give information on the amount of salt contained, on the alkalinity or acidity, and on the presence of organic substances in solution.

The bathythermograph can also be used when the vessel is travelling through the water at full speed. In this way temperature observations can be made quickly over large areas. An elaboration on this now coming into use is an instrument known as a temperature-salinity probe, which adds the measurement of salinity. This records continuously through a cable as it is lowered and raised from a stationary ship. This also measures the pressure, and so the depth. The temperature is measured by a thermistor, and the salinity by the electrical conductivity of the water. As conductivity is also temperature dependent, it is necessary to have a second thermistor in circuit in the salinity meter to supply the correct conductivity reading. All the results may then be fed into a computer which will supply charts of temperature and salinity distribution.

While the water bottle is thus being superseded by modern methods for speed of operation and increased extent of coverage, this older instrument is still necessary. It still has to be used for collecting samples for chemical analysis of nutrients and other constituents, and since laboratory analysis is the most accurate it is necessary to collect samples by water bottle in order to calibrate the modern instruments.

A further physical observation that has to be made is the penetration of light into the sea. It is necessary to know the strength and colour of the light at different depths. A rough and ready way of noting the transparency of the water from place to place is to lower a circular, white disc to the depth at which it disappears. This method of course gives little information beyond a mere rough comparison of the transparency of different regions. Special cameras have been used which can be exposed for a definite time at any required depth, but they have yielded little information beyond recording the actual presence and colour of the light. Nowadays the light is measured electrically by means of photo-electric cells and by this means great advances have been made in our knowledge of the light under the sea.

One of the most recent methods used for studying the world distribution of ocean currents is infra-red photography from space satellites. These photographs show up areas of water of different temperatures, so that the distribution of warm and cold water and of regions of upwelling can be watched, as are also the meteorological conditions.

This completes a short outline of the methods used in studying the physical and chemical conditions in the sea—the depth, the currents, the temperature and so on. It yet remains to be described how knowledge is obtained of the living world present therein. All the various animals and plants that live either in or on the bottom, or swimming or drifting in the water above the bottom, have to be captured and the methods of catching have to be well thought out because not only are the animals wanted in order that their structure may be

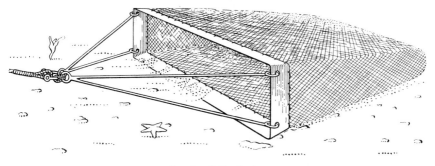

A naturalist's dredge

examined, but also that numbers may be obtained showing their actual abundance from place to place and from season to season.

One of the first instruments to be used for the capture of animals from the sea bottom was the 'naturalist's dredge'. This varies in dimensions according to the size of the ship from which it is to be used, but in essentials it consists of a rectangular or triangular shaped frame of iron to which is attached a short bag of strong netting. The mesh of the net depends on the size of the animals that are to be caught. The frame at the mouth of the net has sharp bevelled edges which can dig through the sand and gravel, and will scrape off any animals that are fixed tight to the rocks. This net is used for all sluggish bottom animals such as starfish, shellfish and sea urchins, and also for those animals that burrow in the surface layers of the sand or mud.

For the swimming animals nets similar to those used by fishermen are employed. These, consisting of trawls, drift nets and seines, will be described in detail in the chapter on the sea fisheries. In addition a small net known as the Agassiz trawl, named after its inventor, is used for deep-sea work. This net has the advantage that whichever way it may fall on the sea bottom it can still fish effectively, a very important point when fishing in great depths, as it is impossible to ensure that the net may not turn over several times in its long journey to the bottom.

A somewhat similar deep-sea trawl is now in general use. This has a weighted shoe which ensures that it will reach the bottom the right way up. The shoe can also be adjusted up or down to vary the distance of the foot-rope from the bottom. This net may be fitted with a sonic device to tell when it has arrived on the bottom, which is otherwise very difficult to know when working in great depths (Plate 34).

By the means already described all sorts of animals may be captured for examination, but the instruments used cannot be said to be quantitative; they will not supply accurate information as to the abundance of the animals on the sea floor.

For this latter purpose instruments known as grabs have been made. One, essentially the same as a coal-grab, is constructed so that when it strikes the

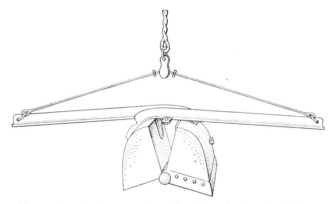

The Van Veen grab, a different type from that shown in Plate 34. (After Holme)

bottom it literally bites out a piece of the sea bed. The instrument is very heavy, and consists of two hinged jaws which are open as it sinks to the bottom; on striking the gravel or mud the jaws sink in under their weight, and as the grab is hauled up they are pulled together so that a solid sample of the sea floor is retained (Plate 34). The sample is then sifted through a series of graded sieves and the sand and mud washed away. All the animals remaining in the sieve can be picked out and counted and so a knowledge of the population on a definite area of the sea floor is gained. It has been shown that this type of grab or bottom-sampler is not completely efficient for studying life on hard, sandy bottoms, as owing to the firm consistency of the sand the grab does not sink deep enough and so fails to catch many of the creatures that live in burrows a few centimetres below the surface. To meet this difficulty another instrument was invented, having a pump which is put into action just when hauling-in commences. Until the pump has finished its work, however, the whole machine cannot be lifted off the bottom. The pump is operated by a wire running over a pulley wheel, and until all the wire is pulled off no strain comes on the instrument itself. With the suction set up by the pump the sampler sucks its way into the sand down to a depth of just over 30 cm, and so a sample of known area and depth is procured. On a similar principle is a little sampler that collects a known volume of the detritus overlying the sea floor with all the microscopic life contained in it and in this case the method of effecting suction is very ingenious and no pump is required. The receptacle into which the sample is to be sucked is sealed with a glass disc. As this is done on the deck of the ship the pressure inside must be equal to that of one atmosphere. At the sea bottom the pressure inside the cylinder is still one atmosphere, but the pressure of the surrounding water is considerably greater. On striking bottom the glass disc is broken by a sliding tube with a point on it and, because of the difference in pressure inside and outside, a sample is forced into the receptacle. There are now a great variety of bottom samplers in use.

It is a more difficult matter to find out how many fish there are in the sea. Fish can swim very fast and so avoid nets; therefore it is not safe to argue, from a catch of fish, that all those present over the area scoured by the net have been captured.

More indirect methods have to be employed and the chief of these is to mark a sample of fish, let them loose, and then find out how many are recaptured. More will be said about the marking of fish under the chapter on fishery research.

There remains the problem of catching the members of that huge drifting community, many of which are so extremely minute and delicate. The instrument *par excellence* of the plankton research worker is the tow-net. This is simply a cone-shaped bag made of silk or other material attached to a ring. On the size of the meshes of the material naturally depends the size of the organisms which will be caught.

If only the larger animals are required, a coarse material will be used so that the smaller creatures will filter through and the catch will consist merely of the kinds of animals that are wanted. The large drifting animals also are more widely distributed in the water so that a larger net will be required than for small plankton. In towing it is essential that as much of the water as possible will be filtered in the net's journey through the water. For this purpose the net itself is made very long, a length three times the diameter of the mouth being the general rule. For the larger animals the net generally used is the ring trawl which has an opening of 1·8 m diameter and is about 5·5 m long; there is thus a tremendous area available for filtering the water. The material in these big nets is hemp.

For the smaller animals the nets are usually about 0·5 m in diameter at the mouth and the material is silk or nylon. The coarseness of the silk depends on the size of the organisms required, the very finest being used for those minute drifting plants, the diatoms. The best kind is that used by millers for grading their flour through, known as bolting cloth, and it has the advantage that the required size mesh can at once be obtained and that these meshes are so constructed that they keep a very constant size.

Much time has been spent in working out how much water the various nets will filter at different speeds of towing, because it then becomes possible to calculate how many creatures are present in a given volume of water. To this end many nets are now fitted with flow meters.

In studying the distribution of plankton organisms in the different water

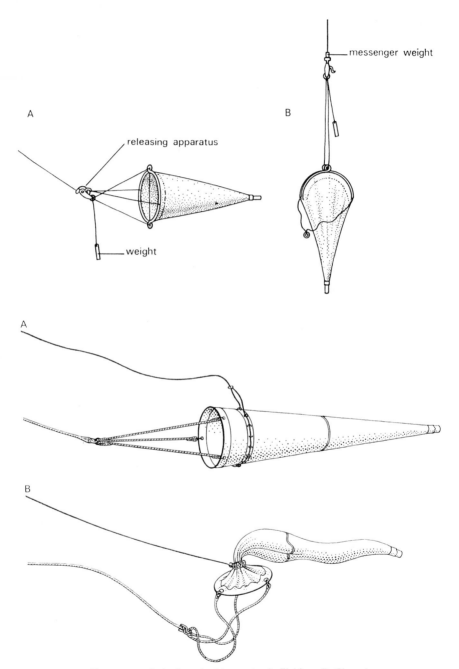

Two types of closing plankton nets. A. Fishing. B. Closed
(Lower figures after Newell)

layers it is necessary that the net should fish only at the required depth and not at all on its journey up and down. For this purpose many different kinds of nets have been constructed that can be sent down with their mouths shut; on arrival at the required depth the net can be opened, and when it has fished for long enough it can again be closed so that it catches nothing as it is being hauled up to the surface. In this way the actual depth at which the different kinds of organisms live can be found.

In order to study the distribution of plankton over as wide an area as possible and to save time at sea, special nets enclosed in streamlined metal cylinders have been designed, which can be towed through the water at high speed. The use of plankton recording instruments for the same purpose has already been referred to on page 92.

But besides the plankton organisms caught by the tow net, there are many forms so minute that they pass through the meshes of the finest silk obtainable. Their presence in the sea was first shown by an examination of the stomach contents of those small planktonic tunicates, the *Oikopleura*, whose houses and fishing apparatus were described on page 134. Many forms were discovered there which had never been seen before in any of the tow-net catches, so that a new method had to be devised to obtain them in sufficient quantity to throw light on their distribution and importance. It was found that this could be done if a sample of water was centrifuged in a small, pointed tube. Samples of water are obtained by means of a pump and rubber hose let down to the desired depth, or, better, in the water bottle mentioned above for the collection of water for chemical examination. The water is then placed in small glass tubes, which taper to a blunt point at the bottom end, and centrifuged for two or three minutes. The surface water is then poured out of the tubes and the little drop that remains at the bottom contains the minute plankton organisms. This drop of water is sucked off with a glass pipette and spread evenly over a glass slide with squares ruled on it which can then be examined under a microscope and the organisms present in a given number of squares counted.

Much interesting work on the geographical distribution of plankton can also be done on ocean liners. If small silk filters be hung under the salt-water tap in the bathroom many of the surface-living animals and plants can be obtained.

One of the most important items in the equipment for oceanographical research is a research ship. On big expeditions a vessel of large size is necessary if the object is to study the physical and biological conditions over extensive areas and at all depths. There must be plenty of space to stow the many and varied instruments and nets. A powered winch capable of holding several thousand metres of strong wire cable has to be used for working the nets at great depths, and probably one or two smaller ones for working small plankton nets, sounding, and collecting water samples. Space should be available on board for working laboratories in which the collections can be sorted and preserved, and water analysed; accommodation should be provided for a small library of the most necessary works of reference for the identification of animals and plants. Large, modern research vessels are equipped with much electronic gear and computers for storing the information gathered.

The costs of equipping and running such ships are heavy and beyond the means of private individuals. Fully equipped research vessels are regularly employed by the governments of many countries for routine research in the areas of the great sea fisheries; many expeditions are also now sent out to study conditions in the open ocean, financed either by governments or research organizations.

Since the second world war a new era has begun with the introduction of submarine research vessels. This was pioneered before the war by Barton and Beebe who made descents far beneath the surface in the bathysphere, suspended on a cable. From this idea Piccard developed the free-moving bathyscaphe, a spherical observation chamber beneath a cylinder containing fluid lighter than water. The bathyscaphe moves under its own power and can rise or sink in the water by releasing weights or fluid, like a balloon. Many direct observations have now been made from this type of submarine vessel, one of which reached the bottom in the deepest part of the ocean (p. 5). A number of deep-water submersibles are now in general use, and in some places underwater laboratories are in operation on the sea floor (Plate 43).

But this is not to say that research cannot be carried out at sea without a large vessel. Much information of the most fundamental importance has been gained by working from small sailing or motor-driven craft. Many of the basic principles underlying marine biological phenomena have been discovered only by exact and minute study within a small area of the sea, and to this end a small boat equipped only with the necessary apparatus for some special branch of research is all-sufficient.

There remains the marine laboratory, accounts of some of which have been given in the first chapter. Their importance in oceanographical research cannot be over-estimated. Much has yet to be done in the way of description of the life-histories of many animals, and the different stages of development have to be carefully drawn in order that they may be recognized when taken in collections by future workers. Under laboratory conditions even the most delicate animals can be reared from the egg and their identification thus becomes certain. Other fundamental problems can be tackled in the laboratory and mention may be made here of one concrete example.

It has been the aim of marine research workers to get as near a true evaluation of the sea as possible. By counting the planktonic organisms in a given volume of water after centrifuging it was found that there were fourteen organisms in a cubic centimetre of water taken from the sea near Plymouth. It was thought that this represented a true picture of the density of population until an ingenious laboratory method for counting the organisms present was devised. In the laboratory half a cubic centimetre of water taken from the sea was added to a large volume of sterilized sea water. This large volume was then divided into seventy parts, each of which was put into a separate flask. These seventy flasks were then allowed to stand for several days, and on examination it was found that 'cultures' of various organisms had grown in them. On the assumption that each culture of each species must have sprung from at least one individual of its kind, which must have been present in the original half cubic centimetre, it was found that the original number of organisms present in the sea water was at least 464 per cubic centimetre—a considerable advance on the fourteen found by the centrifuge method.

In conclusion, it can of course be realized that the introduction of aqualung diving has now made possible direct observations of the habits and abundance of marine organisms beneath the sea surface and most laboratories have their own teams of divers.

Chapter 14
THE SEA FISHERIES

The sea fisheries have changed much over recent years. When the first edition of this book appeared in 1928 it was stated that in 1924 the value of the sea fisheries of Great Britain and Ireland amounted to twenty million pounds and was three times as great as that yielded by the fisheries of any other northern European nation. In 1958 the value, due to increased costs, had risen to nearly fifty-eight million pounds, but by comparison with other European countries the British proportion had dropped considerably, Norway being first with 18·2% of total landings and Great Britain second with 13·7%.

But even more striking is the changing order of world catches. Before 1961 the North Atlantic fisheries were the greatest, but in 1961 the North Pacific fisheries took first place. In 1959 Japan led the world, Russia and the United States coming next; but in 1961 Peru had moved up to second place, and in 1962 she went ahead of Japan, and was still in the lead in 1970.

The total weight of the world catch rose from twenty million tonnes in 1938 to thirty million in 1959, to over fifty-one million in 1964, reached over sixty million in 1968, and nearly seventy million in 1970.

In 1964 Britain was in thirteenth place in the world order, the first five being Peru, Japan, China, Russia and the United States.

Not only has the extent of sea fishing grown enormously, but there have also been great changes in the composition of the catches. In earlier years the catches consisted mainly of fish for human consumption. Now a high proportion of the fish caught are used for making fish meal, especially those from Peru. In 1968 more than one-third of the fish caught were used for these industrial purposes. This large increase in the production of fish meal is required for broiler chickens and for animal husbandry, to which indeed it is due. Another change has been the great increase in fisheries for shrimps which now form a large part of the catches used for human consumption, while squid are caught in great quantities, especially in Japan. In 1969, 1200 boats were employed in Japan which caught 50,000 tonnes. The squid are caught by jigging, and lights and echo sounders are used. It has been estimated that the potential sustained yield of food products from the sea is not much above 100 million tonnes per year. If this is so the limit is already being approached.

Many kinds of fish make up the total catches, but most of these fish are clupeids, that is herring, pilchard, sardines and their relatives, and gadoids, namely fish like cod, haddock and hake, which are mostly northern water fish. In tropical waters most of the food fish are pelagic, belonging either to the clupeids or to the mackerel and tuna family.

British sea fisheries may be divided roughly into deep-sea and inshore fisheries and of these the deep-sea are by far the more valuable. Amongst the inshore fisheries are classed the fishing for crabs, shrimps, lobsters, and mussels, clams and oysters. This, the shellfish industry, is of great economic importance (Chapter 16).

The British deep-sea fisheries other than that for the herring are on the whole carried out far from the land by vessels which may remain away from their port for two or three weeks. In olden days the fishermen, having to rely on sail alone could not venture too far from their harbours, because of the time required to return to port and the consequent difficulty of keeping their catches fresh. But with the advent of steam vessels carrying large supplies of ice the fishermen went farther and farther afield, until they ranged from Greenland and the White Sea in the north, to the Moroccan coast in the south. In consequence the powered ship superseded the sailing vessel, and the picturesque brown-sailed trawlers and smacks are no longer to be seen sailing from Brixham, Plymouth and Lowestoft.

The type of ship used in these deep-sea fisheries is dependent upon the kind of fish that it is required to catch, whether it be fish living upon the sea bottom, demersal fish, or those that roam the water layers above the bottom, pelagic fish. For the former, the bottom fish, the vessel in general use is known as the trawler; for the latter, it used to be the drifter, each type being specially designed and equipped for the kind of gear used to ensnare the fish.

Since the second world war a revolution has occurred in fishing methods and this still continues. The study of fishing methods has become a technological science and all kinds of improvements have been made to existing gear and new methods have been evolved. Extensive experiments have been made with models, and the working of the nets in the sea has been observed by aqualung divers, by underwater photography, and with the help of echo sounders.

Basically all this recent development has sprung from the two established fishing methods of trawling and drifting. A detailed account of these two methods is therefore still useful for an understanding of the historical sequence of the evolution of fishing methods.

Trawling

Trawlers are larger than drifters, and they vary slightly in size according to the grounds they frequent, the vessels required for long voyages to such rough and inhospitable waters as the Arctic seas being the larger and more powerful. In general they range between 36 and 61 m in length, and in recent years in place of steam most have been powered by diesel motor engines, and some by diesel electric.

The following are the main requirements of a trawler: low freeboard to help in the hauling of the net over the side; a flush deck; a powerful winch capable of carrying many hundred metres of heavy wire rope for hauling the huge nets; a well-protected propeller, and a capacious hold for storing ice and fish. The vessels are built of iron and are capable of speeds up to thirteen or more knots. They are fitted with electric light, and radio, and with refrigerating apparatus.

The net used is known as a trawl, of which there are two kinds, the beam-trawl and the otter-trawl. The beam-trawl went out of fashion for a period when

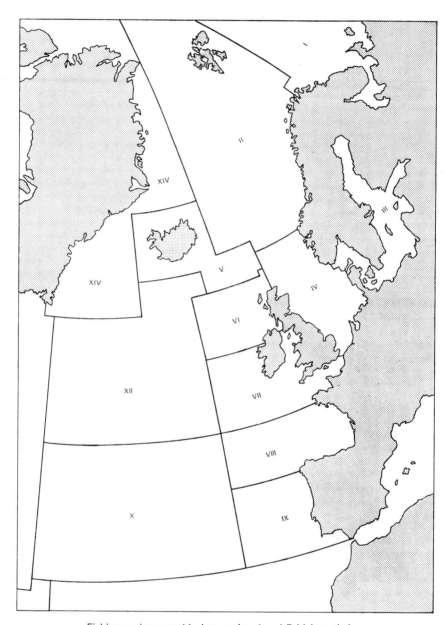

Fishing regions used in International and British statistics

I Barents Sea. II Norwegian Sea, Spitzbergen and Bear Isle. III Kattegat–Baltic. IV North Sea. V Iceland–Faeroe. VI NW Scotland, N Ireland and Rockall. VII English Channel, rish Sea, S and W Ireland, and Porcupine Bank. VIII Bay of Biscay. IX Portuguese waters. X Azores. XII North of Azores. XIV East Greenland.

the otter-trawl took its place. It was still used for catching shrimps and is now being used again for the capture of flat-fish, especially soles, for which purpose it is very efficient. 'Tickler' chains are attached to disturb the buried soles, and electric fishing is being tried in place of chains.

The trawl net is of a flattened conical shape. The top edge of its mouth is straight in the case of the beam-trawl, while its lower edge is in the form of a hollow curve and has running along it a heavy foot-rope; this sweeps the ground as the net is fishing and stirs up the fish which, rising upwards, are swept into the mouth of the net. The net itself tapers away behind, narrowing down until the final 3 m of the net are reached; in this portion, known as the cod-end, or purse, the sides are parallel, and laced through the extremity is a rope, the cod-line, by means of which the purse may be closed by drawing tight (Plate 35). It is in this portion of the net that the fish collect, and on the arrival of the net on board, the contents can be emptied out by merely untying the cod-line. The fish, once having reached the cod-end, are prevented from swimming forward in the net again and possibly escaping out at its mouth, by valve devices known as flappers and pockets. The lower surface of the cod-end, that portion of the net which receives greatest

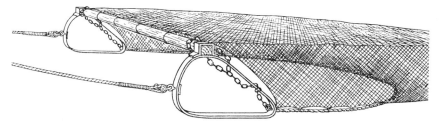

Beam-trawl, showing frame and front portion of net

wear by dragging over the sea bottom, is protected by stout pieces of old netting known as the rubber or false belly, the under-surface of the whole net itself being called the belly. In these main essentials the nets used for either beam- or otter-trawls are much the same, but a difference lies in the means by which the mouth of the net is kept open as it sweeps over the bottom.

In the front of the net in the beam-trawl is a frame, consisting of two D-shaped iron runners, known as shoes or trawl heads, joined together above by a long and very stout wooden beam. To this beam the upper edge of the mouth of the net is attached, the short side of the mouth being fixed to the runners while the heavy foot-rope, curving backwards at the centre, drags along the ground. The opening of the net is therefore determined by the size of this frame, its width being the length of the beam and its height the same as that of the shoes. The length of the beam may vary from 13 to 15 m, down to much smaller dimensions to suit the size of the boat in which it is used. The height of the shoes for a full-sized trawl is about 1 m, so that the largest opening a net could have would be a rectangle 15 m by 1 m. The net itself may extend backwards up to 30 m in length.

The principle upon which the mouth of the net is kept open in the otter-trawl is entirely different. In this case there is no frame to which the front of the net can be attached. Advantage is here taken of the fact that if a kite or any flat object set at an oblique angle be drawn through air or water it will move in a direction outwards. The well-known method of poaching for trout by means of an 'otter' is based on this principle. A flat piece of wood set at an angle is drawn through the water by the poacher on the bank. As he goes forward the 'otter' also goes forward, but owing to the angle at which it is set it moves ever farther out from the bank, towing behind it a string of lures which are thus presented to fish lying far beyond the reach of anyone upon the bank fishing in a sportsmanlike manner.

In the case of the otter-trawl two 'otters', or doors, are used. To these the sides of the net's mouth are attached, and they are set at such angles that as they are drawn over the sea bottom they diverge farther and farther from the centre of the net's mouth until an equilibrium point is reached and the mouth of the net is stretched agape. The upper edge of the mouth of the net is in this case not straight, but like the foot-rope it curves backwards, and the foot-rope is the longer of the two so that the net immediately behind the head-line forms a roof over it. The tremendous strain upon the bag of the net as it moves through the water keeps the mouth of the net open vertically. The otter doors are made of heavy iron-bound wood and are about 2·5 m in length, and 1·2 to 1·5 m high. The lower edge that runs along the sea bottom is heavily encased with iron to form a shoe 7·6 cm thick.

In the case of the beam-trawl we saw that the opening of the net was limited by the size of the frame. Above a certain size the frame becomes too large and cumbersome for practice. With the otter-trawl, on the other hand, the opening that a net may have is limited only by the actual opening of the net itself, the doors being practically the same size for any net, their weight alone being increased slightly for the larger nets. It stands to reason therefore that an otter-trawl can be used with a far larger opening than any beam-trawl, and for this reason it superseded the beam-trawl. The height of the mouth also is not limited by a frame, and in an efficient otter-trawl it is probable that the upper edge of the mouth or head-line may be 3 to 5 m above the bottom, so that the chances of a fish escaping over the top of the trawls are considerably decreased. These trawls are used nowadays with an opening of anything up to 30 m in width.

This bag-shaped net is drawn over the sea bottom by the trawling vessel. The weight of these nets out of water is great and, when this large area of netting is exposed to the friction of the water as it is towed along, the pull is enormous. For hauling these great nets powerful winches are required, carrying on their drums wire cable sufficient to withstand a strain of many tonnes, for when by accident a net comes fast against a rock or submerged wreck the whole weight of the ship is taken by the wire.

Much skill is required for 'shooting' these nets and an inexperienced hand might soon be in difficulties through the net and warp winding round the propeller of the ship, or going down belly upwards.

The net is hauled by two wire warps, one attached to each door (Plate 34). These wires run over stout pulley wheels fixed in two heavy frames known as the gallows, one forward and one aft. By the presence or absence of the gallows one can tell at a glance whether a vessel is fitted for trawling. On most trawlers there are two pairs of gallows so that the trawl may be shot from either side of the ship. Generally two nets are in use, and as soon as one has been hauled up full of fish on the port side, the starboard net is shot and so there is no waste of time while the catch is being emptied. While the starboard net is fishing the catch of the port net is sorted and the net is prepared ready to shoot the moment the other net comes on board. While in most otter-trawls the doors used to be attached directly to the wings of the net, usually nowadays there is inserted between each door and the net itself a pair of warps, the lower of which sweeps along the ground, tending to drive fish inwards. In this way a wider area of the sea bottom is swept, and by using glass floats on the head-line of the net its vertical opening is increased. Recently trawlers have been designed so that the net can be fished over the stern, and this type of trawler is gradually superseding the side trawler (Plate 35). The stern trawler has certain obvious advantages over the side trawler whose design had followed on from the use of the beam-trawl which was easier to haul and shoot over the side. It takes a much shorter time to shoot and haul the trawl over the stern and the trawl can be shot with the vessel steaming head to wind. In addition the fishermen are working in a safer and less exposed position on the deck.

The great trawling grounds of the north-east Atlantic extend from the White Sea in the north, along the coasts of Norway, around Iceland and the Faroe Islands, the north of Scotland, the whole of the North Sea, and the Baltic, in all the waters that bathe the western shores of the British Isles, in the English Channel, the Bay of Biscay, southward along the coast of Portugal to the coast of Morocco. Over all this area the catches vary much in composition, owing to the geographical distribution of the fish. The most important food fish caught in the trawl are the haddock, the cod, the plaice, the whiting, the hake, the coalfish and the dogfish. Other kinds taken in smaller quantities are all flat-fish, such as the sole, the turbot, and the brill, the skates, gurnards and many others. The catches of trawlers are divided into white-fish or round-fish, such as the cod, and flat-fish.

Trawling is chiefly carried out in water down to 200 m in depth, but some of the big trawlers when out for hake on the continental slope of the Atlantic have fished in water as deep as 1000 m, although 400 to 600 is the more usual depth. When fishing over rough ground, round rollers or bobbins may be attached to the foot-rope to prevent it catching in the rocks. With the use of echo sounders mounted on the trawl it is now possible to set the depth so as to catch such fish as cod and haddock above the bottom.

Although, as described below, herring are usually caught near the surface there are certain times when they frequent the bottom and an industry has arisen for catching them in the trawl also. When fishing for herring, however, the mesh of the net is somewhat reduced and the trawl is towed at nearly the full speed of the vessel, as opposed to the two and a half to four knots under usual trawling conditions.

One of the developments of the trawl, resulting from fishing for herring, has been the rearrangement of the otter boards and their substitution by 'kites' so that the net can fish above the bottom. Thus has evolved mid-water trawling for pelagic fish (Plate 34).

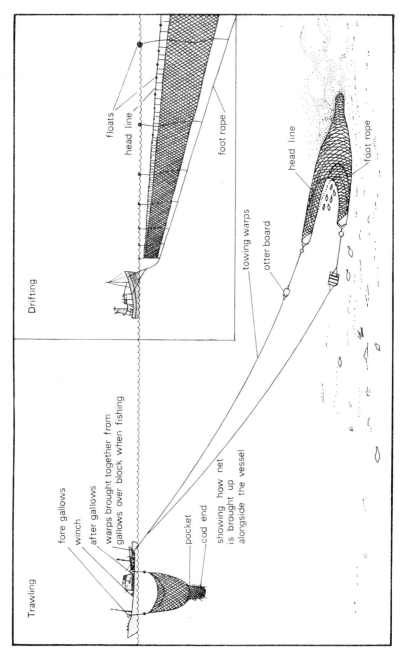

Trawling

fore gallows
winch
after gallows

warps brought together from
gallows over block when fishing

pocket
cod end

showing how net
is brought up
alongside the vessel

towing warps

otter board

head line

foot rope

Drifting

floats

head line

foot rope

head line

foot rope

Methods of trawling and drifting

Drift Fishing

Most of our food fish live on or near the bottom and can be scooped up by the trawl, but two or three species, such as the herring and the mackerel, live a great part of their lives swimming in huge shoals in the water layers above the bottom. In the daytime they are usually swimming rather deep in the water, but at dusk and dawn they come up very close to the surface itself. It is then that they can be most easily caught in the specially designed nets known as drift nets. This habit of coming towards the surface at night is not peculiar to herring and mackerel. Those drifting animals that form the plankton show these habits and such vertical migrations are also common to many other fish. The hake is an excellent example; so regularly does this fish leave the bottom at night that the fishermen never trawl for them except in the daytime, when they are on the bottom.

The drift nets by which the herrings are caught act on a principle very different from that of the trawl. There is no bag into which the fish are swept. A drift net is literally a wall of netting hanging vertically in the water, and in this case it is the mesh of the net that catches the fish. This mesh is made so that the herring can exactly push its head through, but not its body. When once the head has been pushed in beyond the gill-covers it is impossible to get it out again, because, the gill-covers being slightly raised, the twine of which the mesh is made slips under them. The herring, swimming in vast shoals, presumably cannot see this wall of netting in the dark and rush straight into it. Great quantities thus become enmeshed and strangled.

The catching of herring with drift nets has now very largely ceased. The cost of nets is high and the labour involved in hauling the net is great. It was found that the purse seine, which required fewer hands to work, was more efficient. With the use of echo sounders to locate a herring shoal and determine its depth a whole shoal could be enclosed in the net (Plate 43). Herring are also now to a considerable extent caught by trawlers either on the bottom or in mid-water trawls.

Drift nets are however still used extensively for catching salmon and tuna. As drift netting for herring has played so important a part in past years in the development of the fishing industry, the description of this method and handling of the catches given in previous editions is retained here as it has much historical interest. Perhaps, too, should fuel become really scarce this type of fishing might have to be resumed.

The vessel employed in this kind of fishing was known as a drifter, and as with the trawlers this became power driven. The drifter is smaller than the trawler, being about 27 m long, and having a speed of about nine and a half knots. The ships are not usually fitted with large trawling winches, but carry capstans on the forward deck, and the foremast is constructed so that it may be lowered when the vessel is fishing. The nets are generally 31 m long and 12 m deep, and they are strung together so that they may form a wall of netting as much as 5 km in length. The upper edge of the net is buoyed with corks, while the lower edge is generally slightly weighted with lead. Very often the net is not fished actually at the surface but a few metres down; it is then attached at intervals by short ropes to buoys or floats on the surface. When great lengths of net are used a strong foot-rope is

o

attached at intervals by short lengths to the bottom of the net. When this rope is hauled on, it takes much of the strain of the heavy nets which would otherwise tear under the weight of a large catch. On smaller boats, however, the foot-rope may be dispensed with.

At about two hours before dark the fishermen start to shoot their nets, and this they do across the tide.

When the complete length of netting is out, the ship lies with one end attached to it and drifts for several hours with the tide. At about dawn, or even sooner, the nets are hauled over the side by power over special rollers, and the herring shaken out of the meshes as the many metres of net are piled down into the hold.

Since the last war the echo sounder has come into general use for locating shoals of fish and this is especially so in the herring fisheries both for drifting and mid-water trawling.

The herring fisheries lie all round the coasts of the British Isles, and the chief time for catching the fish by drift nets is when they are approaching the coasts in dense shoals to spawn. Time of spawning is probably to a large extent determined by the temperature of the surrounding water and there are great differences in the dates of the spawning periods from place to place. The great drift fishery used to start on the west coast of Scotland at Stornoway in the Hebrides in May; in June the herring were being caught off the Orkneys and Shetlands, after which they began to appear successively at different localities down the Scottish coast, until by the middle of July the fishery opened at Shields.

By the end of July, Scarborough and Grimsby were the chief centres of the herring industry, and early in October the fishermen were hard at work at Yarmouth and Lowestoft, the two leading herring ports. In recent years the catches from these ports have been decreasing. Many Lowestoft boats used to end the herring season fishing from Plymouth, where the herring spawned throughout the winter until January, but the herring ceased to come to Plymouth. While there was still a herring fishery in western waters between the English and Irish coasts, the Plymouth fishery died out in the early 1930s, the herring being replaced by pilchard. This was the result of climatic change (see p. 224) and the herring may return, but the drift net will not now be used to catch them.

The apparent advance of the herring southward along our coasts gave rise to the idea that it was actually a migration of the herring themselves; that in the spring they collected in great armies in the Arctic Ocean and from there worked their way down our coasts. Thomas Pennant in his 'British Zoology' in 1812 remarks, 'The great winter rendezvous of the herring is within the Arctic circle; there they continue many months in order to recruit themselves after the fatigue of spawning, the seas within that space swarming with insect food in a far greater degree than in our warmer latitudes.

'This mighty army begins to put itself in motion in the spring: we distinguish this vast body by that name, for the word herring is derived from the German, *Herr*, "an army", to express their numbers.

'They begin to appear off the Shetland isles in April and May: these are only the forerunners of the grand shoal which comes in June, and their appearance is marked by certain signs, by the number of birds, such as gannets and others, which follow to prey on them: but when the main body approaches, its breadth and depth is such as to alter the very appearance of the ocean. It is divided into distinct columns of five or six miles in length and three or four in breadth, and they drive the water before them with a kind of rippling: sometimes they sink for the space of ten or fifteen minutes; then rise again to the surface, and in bright weather reflect a variety of splendid colours, like a field of the most precious gems, in which, or rather in a much more valuable light, should this stupendous gift of Providence be considered by the inhabitants of the British Isles.'

But it is now certain that this was a mistaken idea and that the herring may stay about in the deep off-shore waters not far removed from the region within which they spawn. It is known also that while many may keep thus within a comparatively short distance of the coasts, others may make considerable journeys even out into open ocean waters. It is found that in adjacent ocean waters they grow faster than near the coasts, and it is possible to trace their movements each year by the rate of growth shown on their scales (see p. 218).

In the great days of the drift-net fishery no fish in the sea were caught in such great numbers as the herring. One boat might catch over 100,000 fish a day and the total catch on such a day for Yarmouth would have been 30,000,000.

This great fishery gave employment to an army of workers on land, chief among which were the Scottish fisher girls. While the fishermen were moving round from port to port following the appearance of the spawning herring in each locality from north to south in turn, the fisher girls followed the fleet on land; they were later replaced by gutting machines. In October, Yarmouth, whose streets were thronged a month before with holiday makers, became crowded anew with these Scottish girls. Here they worked all day cleaning and gutting the herring harvest with razor-sharp knives wielded in their dexterous hands. Like a flash the cut was made and the fish ready for preserving, and so they carried on all day, while others were employed in salting, packing, tending the nets and other multifarious tasks.

The value of herring landed in England and Scotland in the year 1958 was 2·7 million pounds. The total value of the herring fisheries of all the nations of north-western Europe was nearly thirty-seven million pounds, over one-tenth of the value of the whole sea fisheries. Expressed in weight, this means that out of the seven and one-third million tonnes of fish landed by these European countries two and one-third million tonnes consisted of herring, which is roughly equivalent to seventeen thousand, two hundred and fifty million fish. This was over twice the number caught in 1924 and was a measure of the increased intensity of fishing.

Since 1958 fishing for herring has been further intensified. The purse seine and trawl have superseded the drift net, and whole shoals can now be captured with the help of echo location. The herring has been overfished (see p. 218).

Seining

A little time before the second world war a new development took place in the British fisheries in net fishing in offshore waters, the use of the Danish plaice seine. This is a bag-shaped net 9 m in length, each side of which stretches out in front to form a 'wing' 24 to 30 m long. The foot-rope is weighted with lead and the upper rope is buoyed with cork or glass floats to keep the mouth of the net open. Attached to the front end of each wing are many metres of warp. When fishing, one end of the warp of one wing is attached to an anchored buoy. The vessel, which is motor driven, goes down tide paying out this warp until it is all out. The net is now put overboard and the ship sails round in a semi-circle, paying out the other warp until it has returned to the anchored buoy. The two ropes are then taken on board and the net is hauled in by a special winch.

It is found that with this net the fish caught are in much better condition than those caught by trawls. In fact the plaice are generally in quite a lively condition and are kept alive on board in tanks by Danish fishermen, amongst whom the net has been in use for many years.

This seine is also used for catching haddock, but in this case the wings of the net are generally shorter than those of the plaice seines. Sometimes the seine is fished between a pair of vessels.

The pilchard seine was a net used off the Cornish coast for catching pilchard in the days when this fishery flourished. When a shoal was sighted by special watchers or 'hewers' from prominent points on the coast the fishermen rowed round it paying out the net as they went. This was very similar to a drift net, and when the circle was completed the pilchard were enclosed within a wall of netting. This circle of netting was then towed slowly towards the shore until the weighted foot-rope was on the ground. The fish were thus cooped up without any way of escape and the fishermen could remove them at their leisure by means of a smaller seine known as a tuck net. This net was very deep and when the fish were surrounded the bottom edge of the net was drawn towards the surface so as to form a bottom and the pilchard were then scooped out in baskets. When a very large shoal has been encircled in the pilchard seine the process of removing them in the tuck net has been known to take several days. A tuck net is also used in inshore waters for catching sprat.

The purse seine is a net used extensively in America for catching the men-haden, which is a close ally of the herring, and the mackerel. A shoal of fish is surrounded in the same way as with the pilchard seine, but the net is made so that a rope can pull the bottom edge of the net together like the mouth of a purse, completely enclosing the fish in a basket of netting. Very large purse seines are used for catching tuna and salmon in the open ocean. As stated above, in European waters the purse seine has now largely replaced the drift net, although the latter is still used for some fish such as the salmon.

Lining

Fishing with hook and line is a method used extensively in some parts for cod. The cod-fisheries of the Newfoundland Banks are world famous and no more vivid description of the arduous life of the fishermen in those waters can be found than in Rudyard Kipling's *Captains Courageous*.

But the actual method of fishing by hand-line, given in that book, became largely superseded by the use of long lines carrying anything up to 3000 hooks each.

The vessels employed were mostly sturdy sailing craft, carrying eight small rowing boats known as dories. The largest cod-fishing fleet is now Portuguese, consisting of schooners each with fifty or more dories. When the banks are reached the dories are put out to lay their lines. The lines are baited with squid, herring, or capelin, and are each 90 m (50 fathoms) long with about ninety hooks. Nine of these lines are coiled in a tub, and each dory carries four of the tubs. When fishing, all the lines are often joined end to end, making up one long line 3292 m long. The hooks are attached at intervals along the line to short snoods about 0·6 m long. These lines may be set in the morning, when they are hauled up three hours afterwards, or in the afternoon when they are allowed to remain down all night.

The fishing is carried out along the Newfoundland coast, on the Newfoundland Banks and in Greenland waters, in depths between 27 and 234 m.

Although the general practice is now to use these long lines there is still a certain amount of hand-lining done. The hand-line is quite short and has only two hooks on it. They may be used from the dories, or at times from the schooner itself when she is drifting. Each fisherman keeps count of the cod he has caught by cutting out their tongues, and throwing them into a basket to be counted up when fishing stops.

When the fish are all aboard they are cut open and cleaned and packed away in salt.

Long lines very similar to those used in Newfoundland fisheries are also used round the British Isles for cod, haddock, whiting and other fish.

Troll lines with baited hooks or lures are towed at speed for catching tuna and salmon, and tuna are also fished for with stiff rods with hooks on short lines, when the fish are literally snatched out of the water.

Inshore Fisheries

Although somewhat hard to define accurately, the inshore fisheries may be said to involve all those fishermen who make their livelihood around the coasts in motor- or sail-driven craft, or from the shores themselves. The fishermen are rarely at sea for more than one night at a time.

Most of the fishing methods used by the trawlers and drifters can be used by these longshore fishermen, but the gear is of necessity lighter and smaller. Small plaice, whiting and shrimps are caught in beam-trawls, and drift nets are used for herring. It is quite a common practice on the east coast of England for the fishermen to shoot their drift nets close inshore at sunset and leave them to drift along the coast with the tide all night, returning to pick them up in the morning.

Seines are used for pilchard, sprat and mackerel from sandy beaches. The rope attached to one end of the seine is retained on shore, and the men row out to sea, allowing the net to run off the stern of the boat as they go. After a certain distance they turn and, making a semi-circular sweep, return to the shore with the other rope. The seine is then hauled in, the two ropes being gradually brought closer together until the two ends of the net are on shore. The central and deeper part of these seines is generally bag-shaped and the fishermen drive the fish towards the bag by splashing as the net is being pulled up on to the sand.

The ebb and flow of the tides over flat sandy beaches is taken advantage of by the use of stake nets. Netting is attached to a number of stakes set upright in the

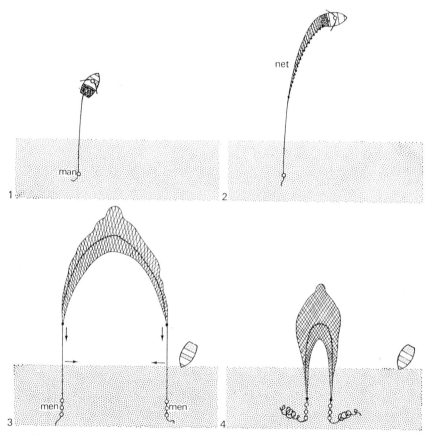

net

man

1

2

men   men

3   4

Diagram to illustrate the principle of fishing from the shore with a seine net

sand to form a trap with a small opening towards the sea which the fish enter on the incoming tide. When the tide recedes the fish are left high and dry on the sand.

In many parts of the world the ancient cast net is still in use. The net is so weighted that a dexterous throw from the shoulder will make it open out and fall flat on the water—a circular disc of netting whose weighted circumference carries it quickly to the bottom with any fish over which it may have fallen. By pulling carefully on a rope the weighted edges can be drawn slowly together over the bottom so that the fish is completely ensnared in the net. Along the Egyptian coast fishermen may be seen using these cast nets for grey mullet which they stalk in the shallow water.

Very often, in the devising of nets, advantage is taken of some peculiarity in the habits of a fish, such as the drift nets which are used for catching the fast swimming shoals of herring or mackerel. The grey mullet is a great jumper and will leap out of the water when it meets an obstruction. If a shoal of grey mullet be surrounded by a seine great numbers may be seen making their escape by leaping over the walls of netting. In the eastern Mediterranean a net is used which besides presenting a vertical wall of netting, has netting lying horizontally on the surface supported on floating bamboos. This net is shot off the stern of a boat which is rowed in a complete circle so that a shoal of fish may be completely surrounded by a hanging curtain of netting (see Plate 38). The mullet immediately jump; but, instead of leaping over the wall as they intended, they find themselves flapping and splashing on the horizontal floating net on the outside. The fishermen row round and pick them up as they jump to their fate.

There is a net used by the Japanese which requires twenty-five to thirty boats and 150 to 200 men to look after it. This is a huge bag net 270 m long with a mouth 76 m wide and 38 m deep; the wings of this net which stretch out on either side in front of the mouth may be as much as 914 m in length. It is used for catching the yellow-tail or amber-fish.

Fish are also trapped, and the gear used varies between the simplest wicker traps and walls of netting arranged in an elaborate manner so as to lead the fish into a closed chamber. A good example of the latter is the madrague which is used along some Mediterranean coasts for catching tunny and may be over 3 km in length.

Hand-lining is a common practice among inshore fishermen, and long lines on the same style as those used in the cod fisheries are also employed. Lines may also be set in the sand at low-water mark which are covered by the rising tide and can be taken up when the water has again receded.

Many fish such as mackerel and pollack can be caught by trailing a shining spinner or feather behind a boat, and one man can operate as many as four lines.

In conclusion it should be said that the scientific study of fish behaviour and capture is leading to radically new methods of fishing. For instance, electric fishing is now practised in some areas, the fish being attracted to a localized position by the electric current and then sucked up in pumps. Another interesting innovation in sheltered waters is the production from a long plastic tube of a curtain of air bubbles rising towards the surface. By this means herring can be diverted into an area in which the nets are operated. The use of surface and submerged lights for attracting fish is also a method commonly practised.

It is known that cod and haddock make grunting noises used for mating or aggression. The time may come when fish will be attracted by the play-back of records of their own sounds! It was of old known that fish could be attracted by sounds, such as by spraying water on the sea surface to resemble the movements of small fish on which they might feed. In this connection the effects of noises made by ships are being studied. It has been noticed for instance that when purse seine fishing for clupeids, mackerel, tuna or saithe the schools were scared by the noise generated by the fishing vessels, especially such noises as those caused by changes in engine revolutions when near the school. Herring appeared not to be scared by a trawler when steaming, but were if the ship was towing a trawl. So it may be that fishing vessels will have to be designed to be as silent as possible.

Some Products from Fish

Apart from the supply of fresh fish for food, some fish are processed in different forms by smoking, salting or drying to form such table foods as kippers etc. In northern latitudes the herring is smoked more than any other fish, but quantities of haddock and cod are similarly treated. Drying is mostly practised in warm and tropical countries where the sun's heat is sufficient to dry them hard reasonably quickly. Oily fish such as sardines, which are the young of the pilchard, are tinned in olive oil, and other fish commonly sold in tins are tuna and herring. Fish may be filleted and packed frozen for sale from deep freezers and most of these types of processed fish will be well known to readers.

There are, however, other products of which fish meal forms the greatest bulk. This is used as food in the broiler industry and as food for farm animals. Some idea of the importance of this industry can be gained when it is realized that in 1968 of the 65 million tonnes of fish caught about 24 million tonnes were used for making fish meal. Most of the fish used for this so called 'industrial fishery' were anchoveta from Peru, menhaden from America, and herring and sand eels from the North Sea. The fish is cooked, pressed and dried, yielding also fish oil. The heads and bones from filleted fish are similarly made into fish meal.

Whale Fisheries

Whale hunting has been practised for many centuries, but probably the first real whaling industry was founded by the Basques of southern France and northern Spain. This fishery was situated in the Bay of Biscay, and was started in the tenth century when the Atlantic right whale, or Biscay whale, was the object of pursuit. From the tenth to the sixteenth centuries the fishery was centred round the towns of Bayonne, Biarritz, St. Jean de Luz, and San Sebastian and from here the rest of Europe was supplied with whalebone and oil.

The whole history of the whale fisheries has been one long story of extinction through man's want of thought for the future, and by the sixteenth century the Biscay whale was becoming very scarce, and possibly shyer, and the Basques had to go farther afield in their chase, often voyaging as far as the coasts of New-foundland.

About the sixteenth century the existence of the Greenland whale became known, and Europeans turned their attention to the northern waters that it in-habited. This gave rise to the great Spitzbergen fishery which was exploited both by the British and the Dutch. These whales were very abundant, and apparently at first tame and easy to capture; as a result the slaughter was wholesale, and by the end of the nineteenth century the Greenland whale was practically exter-minated, so that these fisheries exist no longer.

At the present day the northern fisheries are practically confined to the capture of the rorqual, or fin whale, by the Norwegians. This is a very fast swimming whale and it was not until the invention of the harpoon gun in 1866 that any serious attempt at capturing it was possible. Owing to its great vitality the old method of capture by hand harpoon was practically useless and extremely dangerous for the fisherman. But while in the first years of the twentieth century these northern fisheries were the most important in the Atlantic Ocean, now the chief hunting grounds are in the frozen south. The whales chiefly taken in the Antarctic are the blue whale, the fin whale and the humpback.

Here, as in the north, the fishery was at first mainly carried out by the Norwegians, but soon the British joined in, after which the Japanese and Russians became equal competitors. The Japanese now have the largest fleet of whaling ships, using whale meat for food. In the period between the wars there were indications of a decline in the stocks and the humpback whale, which in 1910 formed a very high percentage of the catches, was largely supserseded by the blue whale. As a result, the Falkland Islands Dependencies, around whom the most productive waters were then centred, financed an expedition in 1924 to study the life-histories of whales. These *Discovery* Investigations (see p. 9) became a prolonged scientific survey of the whaling grounds from 1925 until after the war. Much new knowledge was gained on the distribution and migration of whales, and on their growth and age at maturity, all of which was necessary for future legislation. In the course of the cruises the conditions of the whales' environment and the distribution of their food were also studied, and great advances were made in our knowledge of the oceanography of the Southern Ocean and Antarctic waters. Now, the whale fisheries are supervised by an International Whaling Commission and close times and overall limits to the catches decided for each year. Despite warnings it is however feared that the blue and hump-back whales are nearing extinction.

In the early days of whale fishing the chase must have been fraught with danger and excitement, for the hunters ventured forth in frail boats and came to close quarters with the whale in order to drive their harpoons in by hand. A whaling vessel usually carried six of these small chasers, sailing boats 8·2 m (27 ft) in length.

Today the whale chaser is a power-driven steel ship of anything up to 40 m in length and capable of speeds of thirteen to fifteen knots. The harpoon is shot from a gun and is itself 1·8 m long and 45 kg in weight. It has four barbs, each 30 cm long, which spring out when the harpoon is buried in the whale's flesh; in the nose of the harpoon is a small shell loaded with powder which is exploded by time fuse and helps to hasten the end of the whale. Attached to the harpoon are 1800 m of stout rope on which, with the help of a steam winch, the whale is played.

The captured whale, with air pumped into it to make it float lightly, is towed back to harbour where the factory is situated and there the blubber is stripped off it and other parts are put through the various processes necessary for producing oil. The whalebone is cut off and the meat is cut into chunks and packed in boxes ready for shipment.

In order to remove the blubber, incisions are made along the whole length of the body. A stout hook is then inserted into the blubber at the head end and by means of a chain and winch the blubber is stripped from the body, while men stand by with sharp knives to part it cleanly from the meat.

When fishing in distant regions where it is not possible to establish a shore factory, large vessels are used capable of performing the final operations on board. The ships generally anchor in some sheltered bay and form a base for their attendant whale catchers to work from; the whales are here drawn up on to the shore to be cut up.

Some ships are fitted with a slipway through which the whole whale may be hauled through the stern and stripped and cut up on board. Such ships can do all the factory work at sea and follow their own whale catchers from place to place.

Seal Fisheries

Almost all species of seals are of commercial value for their hides, oil and meat, but the most important by far are the sea bears or fur seals whose warm furs maintain the sealskin industry.

Like the whale, seals have been wantonly massacred by the hand of man and in many parts of the world they have been hunted to the verge of extinction. At one time the great herds of the Antarctic exceeded those in any other region, but now they have been reduced to small numbers.

At the present day the chief sealing grounds are in the Bering Sea on the Pribilov and the Commander Islands. Although originally these seal herds which numbered several million individuals suffered considerable depletions, the Russians by whom they were exploited showed great forethought. Knowing that the seal was a polygamous creature they took great care to save the female population and killed off only the superfluous males. These seal fisheries, which passed into the hands of the United States of America through the sale of Alaska by the Russians, are now controlled by the American Bureau of Fisheries. Killing is confined only to three-year-old males and a large number of this class are conserved for breeding purposes. When the seals are all assembled on the land for breeding each male gathers round him a harem, averaging about thirty wives. When all the harems are collected the remaining superfluous males or bachelors are kept away from the breeding grounds by the married males and so can be easily herded together and driven off to the killing grounds by the sealers. The killing grounds are situated some distance from the breeding area so as to prevent disturbance there. When the seals have been chosen for slaughter they are knocked unconscious by a blow on the head with a heavy club and rapidly bleed to death while in this condition.

Towards the end of the nineteenth century these seal herds ran great risk of extermination from the practice of pelagic sealing. Advantage was taken of the migratory habits of the seal and, when in the spring the herds were returning northwards along the Californian coast to the Pribilov Islands, men went out in sailing ships to shoot them. In this way great harm was done, for not only was there wanton destruction of females but the death of every female meant the death of at least one pup for future years. After protracted discussions this pelagic method of sealing was prohibited in 1911, and it is satisfactory to know that under the care of the American government the seal population is now nearly as great as it can be (see also p. 256).

Chapter 15
FISHERY RESEARCH

To a naturalist the engrossing work of unravelling the many mysteries of life in the sea and the quest for information to enable a clear picture to be gained of the life stories of the animals is all sufficient. But in these days of applied science the naturalist is also called upon to utilize his knowledge to help the many industries which are dependent on the products of the sea. By far the most important of these products are those fish that are eaten by man—under the term edible fish are generally included both the true fish and the shellfish which consist of crabs, shrimps and oysters etc.

There can be no hard and fast line between oceanography and fishery research, because all of the widely ranging problems attacked by the oceanographer will be found to have some bearing on the lives of the fishes living in the sea. Fishery research is, rather, a branch of oceanography, and its main concern is to attack the problems directly connected with those fish which form the chief supplies of food from the sea.

The foundations for the application of science to the fishing industry must lie in a complete and intimate knowledge of the life-history and habits of each fish. It will perhaps enable the reader best to grasp the methods and aims of modern fishery research if we describe how our present knowledge has been gained of two fish, such as the herring and the plaice, which differ widely in their habits. Both of these fish are of considerable importance to the fishing industry, the herring, indeed, being one of the most important of all fish from the waters on the north-eastern part of the Atlantic. The plaice make up a large part of the catches of the trawlers in the North Sea while the pursuit of the herring constituted that great drift-net fishery which took place around British coasts chiefly in the late autumn and winter. This distinction points at once to a fundamental difference in the behaviour of these two fish. The plaice which are captured by the trawls that scour the sea floor are bottom-living or demersal fish, while the herring, as their mode of capture suggests, live, for considerable periods at any rate, in the water layers above the bottom and are pelagic.

In studying the life-history of a fish it is best to start right from the beginning and watch the growth and movements of the individual from the time it is first brought forth by the parent fish as an egg.

Let us, therefore, put ourselves in the place of the early naturalists, the founders of fishery research, and imagine that we must set out to find the eggs of the herring and the plaice. We should immediately say, 'How can we find them if we do not know what they are like to look at?' This brings us to the very first step in the study of the life of a fish. The eggs when discovered must be accurately described and figured so that those who come after us may know at once what to look for.

Towards the end of the nineteenth century a controversy arose as to whether the advent of the steam trawler might not have a harmful effect on the sea fisheries by destroying in large quantities the eggs of the fish as the heavy net was dragged over the bottom. In the year 1864 the Norwegian naturalist, Prof. G. O. Sars, discovered large numbers of eggs drifting in the surface water layers well above the bottom, and was able to show that these were the eggs of cod. This discovery stimulated the search for the eggs of other fish and it has since been found that, with only one exception, all our food fish have these drifting eggs. This early piece of fishery research immediately showed its practical value; for, when the discussion arose about the necessity of putting restrictions on steam-trawling because of the supposed destruction of eggs, the scientific advisers were able to show that no such danger existed.

It has been said that there was one exception to the rule that our food fish have pelagic eggs. This exception is the herring, whose eggs (Plate 39) stick in clumps to the bottom (see p. 55). In this case, however, there is little risk of their destruction by the trawls, because the eggs lie well down in the crevices between the stones on the rough ground on which they are laid, apart from the fact that the rough grounds on which the herring spawn are so rocky and strewn with boulders that they are unsuitable for trawling.

The eggs of the plaice (Plate 39) and the herring having been accurately described and identified by means of artificial fertilization and by hatching, we can now proceed in our search for the spawning grounds of these two fish. There are two obvious ways of searching for these localities, firstly by seeking for the eggs themselves and secondly by finding from the statistics of the fishing industry at what places most of the fish containing ripe roes were caught. The last method obviously involves the least trouble (given the necessary statistics), but it has the disadvantage that the results are only approximate, as there is no evidence that the fish would have actually spawned where they were caught since they can swim a considerable distance in a short space of time.

Let us then consider the first method. Having roughly marked out the spawning region by means of the statistical method, it is necessary now to put to sea to these grounds. The plaice egg we have said is pelagic and drifts about in the water layers above the bottom. To catch these eggs, therefore, a tow-net must be used. The area to be searched is divided up on the chart into a number of small sections, in each of which a point or 'station' is marked, at which hauls with the tow-net are to be made. Generally, when a research vessel is on a 'plaice-egg cruise', a vertical tow-net is used; that is, a net which is hauled up vertically through the water from bottom to surface. The depths being known, it thus becomes possible to compare with accuracy the numbers of eggs caught in different localities. At the end of the cruise the numbers of eggs taken at each station are marked out on the chart and contours drawn through the stations where roughly the same numbers of eggs were caught (see chart on p. 212). In this way, if the whole area in which the eggs were present has been studied, a map should be produced showing the largest number of eggs caught near a centre, the numbers becoming less and less the farther one goes from this centre. Now the worker, who was counting the eggs, at the same time made notes as to the stage of development of the young fish within the egg. By the use of drift bottles and current meters, as described in a previous chapter, a rough measure can be obtained of the movements of the water in the regions examined. It has been worked out how long it takes a plaice egg to develop under different temperature conditions, and the temperature of the water being known the investigator can

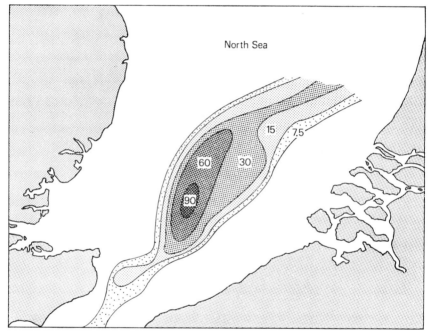

Chart showing distribution of plaice eggs in the North Sea in December, 1910. The figures show the numbers of eggs per square metre surface. (After Buchanan–Wollaston)

immediately reckon how many days have elapsed since the actual spawning of the eggs which he has taken. Knowing also the rate and direction of the water movements he can lay his finger on the spawning grounds.

Other methods need to be resorted to in a search for the spawning grounds of the herring, since its eggs are laid on the bottom. In this case, besides gaining information from the actual catches of herring an examination of the stomach contents of other fish becomes advantageous. It is known that the haddock will greedily devour herrings' eggs, and the occurrence in the trawl of so-called 'spawny haddock' crammed with herring spawn is a sure indication that the spawning grounds are not far off. A search for the eggs is far less fruitful, owing to the fact that they lie in the crevices between stones and are rarely taken in the trawl. On occasion they have been searched for by means of the grab mentioned on page 186, but here again the chances are small of the instrument picking up a sample of the bottom from exactly the right spot. More recently they have been located and their abundance estimated by submarine television and photography.

Let us now trace the life story of the plaice in the southern North Sea. Research carried out on the lines indicated above shows that the mature plaice gather into a region some little way off the Dutch coast: here the eggs are discharged and drift freely in the water for about a fortnight, at the end of which time they have been carried by currents farther north and considerably nearer the coast. From the eggs hatch those little symmetrical larvae mentioned in a previous chapter (Plate 39). The movements of the plaice can be followed by making catches with a very large coarse-meshed tow-net. It will be found that by the time they are about 1·5 cm in length they are being drifted very near to the coast. When they have reached this size they suddenly disappear from the tow-net catches. Their metamorphosis has taken place, that is the left eye has moved over to the right side of the head and the fish has assumed the typical adult characters. At this stage the fish seeks the bottom. It must then be sought with the aid of a fine-meshed trawl, such as the shrimp-trawl. A search will reveal that the young plaice have moved into the shallow water covering the sandy flats that occur all along the Dutch coast. Here fish between 2·5 and 5 cm in length can be caught in immense numbers. When the plaice are of this age they can often be found in the sandy pools left uncovered by the tide. As an illustration of this a photograph is given in Plate 42 of such a pool in the Gannel Estuary at Newquay, Cornwall. Two of us fishing in this small pool for just under an hour with small prawning nets early in June captured the forty young plaice shown in their natural size on Plate 42. There were probably still many left in the pool as the meshes of the nets used were rather large and the smaller specimens could slip through with ease. Thus we see how the coastal waters of Holland can form a nursery ground for a very great number of the stock of plaice in the North Sea. This discovery is of great economic importance. It is on the survival of these fish that the future stock of large plaice depends. Immense damage is, however, done on these grounds by longshore fishermen seeking other fish. So important an aspect has this assumed that it has been seriously urged that during the time the young plaice are there the grounds should be completely closed to trawling.

From a size of 10 to 13 cm onwards we are dependent on the statistics of the landings of fish for information as to their general distribution. But while these figures will give information as to their distribution they tell us nothing of the movements of the fish or of their subsequent growth. It is here that the system of marking the plaice has proved invaluable. A small piece of silver wire is pushed through the back of the fish and on either end of it are attached small plastic discs bearing a number. The fish thus marked are then measured, and their length, together with date and place of capture, is inserted in a book opposite the number corresponding with that on their discs. The fish are then liberated.

Fishermen are notified that for each marked fish returned to headquarters they will receive the market price of the fish and a reward for the disc. When a fish is recaptured it is once more measured and, on comparing its length and date and place of recapture with the original details for the fish bearing that number, one can immediately tell how much it has grown and how far it has travelled since the day it was marked.

In order to study the population of plaice in a large area like the North Sea it is necessary to determine the age of great numbers of fish. Fortunately the plaice carries its own age indelibly written upon it. Situated in a cavity just behind the brain are two small white bones. These lie loose in the ear sac and enable the fish to regulate its balance in the water, just as the semi-circular canals in our ears provide us unconsciously with a means of keeping ourselves upright

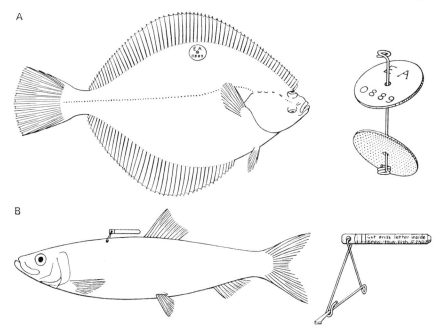

Diagram showing type of marking tag used and position of insertion for A, plaice and B, herring

(Plate 42). Examination of these ear bones, or otoliths, shows that they possess alternate light and dark concentric rings, similar in a way to those on a cross section of a tree trunk. It has been proved by a study of the ear bones of marked fish that these rings correspond to the summer and winter growth of the fish.

Of great importance in the study of the life-history of a fish is a knowledge of its food and feeding habits. The method of finding out the chief food eaten is a very simple matter and merely consists in opening the stomach and examining the contents. When a very large number of fish have been thus examined over a period of two or three years a knowledge is gained of the most common kind of food for the different sized fish and of any changes in the diet that may occur during the seasons. Having discovered the food the next problem is to find out what its distribution is in the sea, because where the food is there we should naturally expect to find the fish that feed on it.

Now, the adult plaice is a bottom feeder; that is, it makes its meals of animals that mostly live on the sea floor and while very many animals are included in its diet, the most important item by far for the adult plaice in the North Sea is a small shellfish known as *Spisula*. To examine the distribution of this sort of food the grab is used and the actual numbers of animals caught in a given area counted, so that, by making observations at stated intervals over a large region, a chart may be drawn up showing the relative frequency of the animals from one spot to another. Such research has been carried out in many localities and it has been found that compared with other places the Dogger Bank is surprisingly rich in these shellfish (see p. 46). The effect of this abundant supply of food has been shown in a very striking manner. A large number of small plaice were caught off the Dutch coast and marked. Of these one half were liberated again at the spot at which they were caught, but the other half were kept on board the research ship in special tanks and transported up to the Dogger Bank, where they were allowed to go free. After a time these marked fish began to be returned

Spisula, the chief food of the plaice on the Dogger Bank. ($\times$ *ca.* 3)

by the fishermen. Examination revealed the fact that those that had been liberated on the Dogger Bank grew much faster than those that had been returned to their place of capture off the Dutch coast. This great growth is to be correlated with the great amount of food present on the Dogger grounds compared with that of those off the Dutch coast.

These results have suggested that it might be a practical proposition to transplant large numbers of fish and so increase the resources of the North Sea. There are unfortunately one or two obstacles in the way, such as a decision as to who should bear the cost of the transplantation, and the expense necessary for patrolling the grounds immediately afterwards to prevent too eager fishermen from at once fishing up all the newly arrived plaice.

One of the most instructive results of these and similar marking experiments was to draw attention to the grave risk of depleting the stock of plaice in the North Sea. Within a year of liberation over a quarter of the marked fish were recaught, 250 to 300 out of every 1000 marked plaice being taken. This may be taken to mean that a quarter of the catchable plaice were being taken out each year—that is those fish which were large enough to be retained by the meshes of the net. At first sight one does not think this to be really possible.

P

We say, the sea is so vast that the efforts of man can have no effect upon the fish population. But when we realize that a fish like the plaice inhabits only the comparatively shallow water, and that this water is of a very limited extent compared with the deep oceans, and when we are told that the area covered by the nets of all British steam trawlers in the year 1926 was about 108,000 square miles (280,800 km^2), and the area of the North Sea itself being only about 130,000 square miles (338,000 km^2), we realize that the chances in favour of the plaice are small indeed. Especially is this so since the trawling area is not distributed all over the North Sea, but is concentrated on certain grounds where the plaice are most abundant. It is such knowledge which has led to the introduction of mesh regulations in the fishing industry.

Let us now turn for a while from the problems of the plaice and follow the study of the life-history of that pelagic fish, the herring. Earlier in this chapter we have described the general methods of locating the spawning grounds of the herring, by noting the presence of ripe fish, by searching for eggs on the bottom with the grab, and by seeking for 'spawny' haddock.

A study of the distribution of these spawning grounds shows that the herring choose out regions in the close vicinity of the coast and of the estuaries of rivers.

It is next necessary to follow the movement of the young herring as soon as they have hatched from their demersal eggs. The baby fish very soon after hatching make their way up from the bottom into the water layers nearer the surface, and here they may be caught by means of tow-nets in the same way as the young plaice.

But when the herring is about three or four months old it has reached a length of about 3·8 cm and can swim too rapidly to be captured by the tow-net. At this stage the fish congregate in shoals and their movements become very hard to follow. We have no really efficient instrument for catching them in the open sea, but they may be caught by fishing off sandy shores and in estuaries by means of very fine-meshed seine nets. In fact research points rather to the fact that it may not be necessary to search in the open sea for the herring at this stage in its life-history, for it is beginning to be realized that the numbers present in the estuaries are very great and probably quite sufficient to give rise to the large supplies of adult herring caught by the fishermen in the open sea. At certain seasons of the year, for instance, the whitebait that abound in the Thames estuary consist almost solely of young herring. When the herring reaches a length of 8 to 10 cm it probably leaves the estuaries, and little is as yet known of its movements until it grows to a size large enough for capture by the nets. In the case of quick-growing fish probably the end of the second year is the time at which they are large enough to appear for the first time in the net catches; the fish would then be about 15 cm in length. But more usually the herring are not large enough to be taken in the drift nets until they are three years old, or at any rate in their third year.

Since the second world war a large trawl fishery for young herring has been developed in the North Sea, chiefly by Germany and Denmark. These are immature fish in their first and second years. This is regarded as an 'industrial fishery' for the catches are used entirely for processing into fish meal. Although it is undesirable that there should be so large a fishery for the small, immature fish it has made possible further research into this intermediate stage in the life-history of the herring.

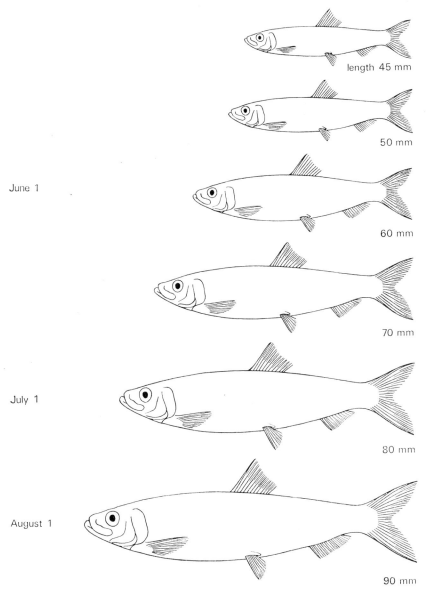

Young herring showing rate of growth during the first year of life. (Actual size)
Born in December to January

In recent years a satisfactory method of marking herring with light celluloid or plastic tags has been evolved. Large-scale marking of adult fish has been done regularly now for some time, and this has thrown much light on the movements and migrations of the adult. The fear that the fishery for immature herring may be damaging the stocks has led to the need for special investigations and as a result large numbers of these small trawl-caught herring are now also being marked.

We have mentioned previously (p. 200) the former belief that after spawning all the herring retired to the Arctic Ocean to recuperate, whence they once more moved down along our coasts in the following year to spawn. This is now known to be an incorrect view, and when not spawning the herring congregate in the deeper off-shore water.

It is found that herring coming from widely separated localities exhibit certain structural and growth differences. Notably is this so in the numbers of vertebrae in their backbones. It is usually found that the herring from coastal areas, where the water is of low salinity, have fewer vertebrae on average than those from more saline ocean waters. Thus Baltic herring have an average of between 55 and 56 vertebrae, those from the western end of the English Channel 56 to 57, and the Iceland herring between 57 and 58. Fishery scientists now for instance have major groupings for herring such as Atlantic herring and North Sea herring. The former may be subdivided into Atlanto-Scandian covering the area Iceland—Faroe—Shetland—Norway, and the latter, as its name implies, Ireland and west and north Scotland. The North Sea herring are subdivided into three main spawning groups: 1. Northern North Sea, or Buchan, whose spawning grounds are situated in suitable places between the Orkneys in the north and the Northumberland coast in the south. 2. Central North Sea, or Dogger, or Bank, herring with spawning grounds from off Shields to the coasts of East Anglia and on the Dogger Bank. 3. The Southern Bight or Downs spawning population whose grounds are in the southern North Sea, the Straits of Dover and the eastern end of the English Channel. Of these three groups the northernmost are summer spawners running into autumn, the middle group are autumn spawners, and the southernmost are winter spawners. In each spawning group spawning tends to start at the northern end first and spread southwards. After spawning, certainly the Buchan and the Dogger herring move east and north-east respectively on feeding migrations to waters off Norway and the Skagerrak. Some of the different populations may thus mingle during their feeding period, but they separate again to spawn.

When counting the herring vertebræ it is usual to bring the fish up to boiling point in water so that it is just beginning to be cooked. The flesh may then be scraped easily from the backbone, without the backbone itself falling to pieces. A much more rapid and clean method which may replace the older method, is merely to take X-ray photographs of the herring, when the vertebræ can be counted on the negative.

The scale of the herring is well suited for studying its growth rate, for the yearly rings containing broad and narrow bands, corresponding to periods of fast and slow growth, are clearly marked out on it (Plate 42). It is known that the scale grows in proportion at a corresponding rate to the body of the fish, and from the distance from the centre of the scale to the outer edge of each yearly ring can be calculated the size of the fish at the end of each year.

Examination of scales from large numbers of herring during the course of many years has thrown very interesting light on the age composition of the shoals of fish. It may, for instance, be found that in one year over 70% of the herring caught were six years old, the remainder being mostly four, five, seven and eight. The fact that these six-year-old fish absolutely outnumber the fish of all other ages indicates that six years before there was a great survival of young herring, far and above those born in the years immediately previous and suc-

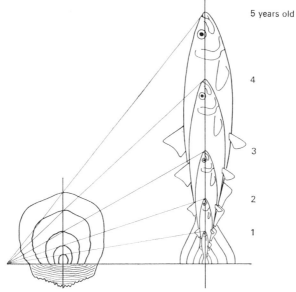

Diagram showing how the scale growth is proportional to the growth in length of the whole fish

ceeding. That year is then known as a good survival year, and it may be found that for several years the catches will be dominated by fish born then.

For many consecutive years Norwegian investigators have studied the age composition of their shoals, and the domination of the catches by a single-year-class derived from a good survival year is clearly shown by the following instance. In the year 1908 the 1904 year-class (i.e. fish born in 1904) appeared in large quantities in the catches. From that year until 1919 these 1904 fish were still predominant, by which time they were fifteen years of age. For eight years, 1909 to 1916, they had in fact practically supported the herring fishery industry; but in 1919 there appeared in the catches a new set of fish hatched in 1914, that is six years old, which nearly equalled the 1904 fish in numbers. Not until 1924 did the 1904 fish die out from the catches, and about this time the 1914 class was also on the wane, but in 1922 a new large brood appeared which was still predominating in 1926. In more recent years good broods have appeared in 1957 and 1961.

Thus we are faced with the amazing conclusion that the main stock of herrings on which to draw may consist for several years of the production of a particular year's spawning, and the outlook for the herring fisheries will be serious if this stock is depleted before there has been another good survival year to keep the supply up to the demand.

It is one of the main aims of fishery research to find out what is the exact combination of conditions necessary to bring about a good survival year, and this knowledge can only be gained by taking very full and detailed observations over a long period of years. With such knowledge it would be possible to predict three years in advance when another large influx into the herring population would occur, because, as has been said before, it is generally three years before the herring has grown large enough to be captured by the nets employed by the fishermen.

This phenomenon has also been shown to hold good for other fish such as the haddock and the plaice. Indeed, it is only reasonable to suppose that it is the rule rather than the exception for each kind of fish to have both good and bad survival years.

In the year 1923 unusually large numbers of baby haddock appeared in the catches of the ring-trawl made by the scientific staff of the Scottish Fishery Board. In consequence of this it was boldly predicted that in three or four years' time there would be a very flourishing fishery. The prophecy was fully borne out. A similar occurrence has been repeated in recent years. It is clear that such predictions must prove of value to the fishing industry as they can obviate an unnecessary glut on the market in the event of a coming good year, or direct attention to some more profitable fish before the lean year comes.

After making detailed observations of the temperature and saltness of the waters of the Danish coasts throughout a period of many years, the Danish fishery investigators discovered a connection between the hydrographical conditions of the water and the mackerel fishery. As soon as the observations of the salinity of the water are known for March and April it becomes possible to predict whether the mackerel fishery taking place two months ahead will be good or bad. The presence of mackerel in this case is probably determined by the strength of certain currents flowing into the Baltic.

It is unlikely that, as yet, such a method will be possible in such waters as those of the English Channel. Conditions in the Danish waters are somewhat special owing to the admixture of fresh water from the Baltic Sea and saltier water from the North Sea. Differences in the hydrographical conditions of the various water layers are there much more marked than we should expect to find in more open water such as the English Channel, where connections between these conditions and the movements of fishes are harder to find. This is not to say that they may not be found some day when our observations are more complete and the methods of making them more perfected.

Now a fisherman is essentially a gatherer of a harvest he has not sown. He goes out and catches fish from the sea but does nothing to increase the supply

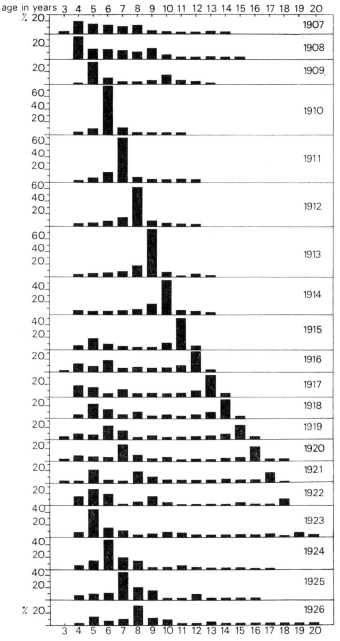

Diagram showing the percentage age composition of the herring caught off Norway each year from 1907 to 1926. The great preponderance of fish born in 1904 in the catches for many years in succession is clearly shown. (After Einar Lea)

contained therein. What has happened in other cases where man has hunted without thought for the future? Whole seal populations have been wiped out; several species of whales have come dangerously near complete extermination and what were once flourishing whale fisheries practically no longer exist. And it was certain that sooner or later there would be a risk of depleting the stock of some kinds of fish. It has been mentioned previously that the plaice, for instance, was apparently being fished out of the North Sea to a dangerous extent.

What might be regarded as one of the largest experiments ever carried out was a direct result of the first world war. For five years the North Sea was virtually closed to the steam trawlers; anyone who went out to fish did so at his own risk. Consequently it can be said that the great plaice grounds of the North Sea were almost completely rested. This afforded a real opportunity to see whether man's efforts as a fisherman could actually have any effect on the stock of fish present in so vast an area of water.

The statistics of the landings of plaice at the big fishing ports in the years right up to the beginning of the war were known. The landings in the years immediately following the end of the five years' rest brought about by the war were watched with the very greatest interest. Would the catches be still exactly the same as in the years before the war, or would the stock of plaice have become recuperated and so afford undeniable evidence of the effects of fishing? The findings were instructive indeed.

In the first year after the war the catches were very much greater than in the years before the war. The catches were also characterized by the presence of very great numbers of old, poorly nourished fish, the grounds being overcrowded. But in the succeeding years the numbers of plaice caught became less and less until they were once more down to, or even below, pre-war figures. The average size of the fish too was much smaller than that of those caught the first year after the war, showing that the fish were being removed at such a rate that they were never allowed to reach a good size. These results may be taken as direct evidence that fishing was being carried on to a dangerously heavy degree. The beneficial effects of fishing up to a certain extent, however, were shown by the fact that after the first year, when large numbers of the old plaice were removed, the growth rate temporarily improved. Similar effects were found after the second world war.

The first care then of fishery research is what is known as a rational exploitation of the sea. That is, that the areas concerned shall be fished right up to their limit, without overstepping it and harming the stock for future years or upsetting too severely the balance of nature. To discover these limits the productivity of the sea must be estimated; the life-histories of all the fishes, and animals and plants that concern them, must be worked out; the causes of the various fluctuations that occur from year to year in the supply of different fish must be found and these fluctuations must be discriminated from those that may have been brought about by overfishing.

It is possible to estimate the total population of some kinds of fish by two methods. One is to mark a large number of fish, liberate them, and see how many marks are returned by the commercial fishermen during the year. If a quarter of them are returned then it is a fair assumption that at least a quarter of the total catchable population has been taken during the year. From the statistics of the total catches a minimal estimate of the total population in the area of sea concerned can then be made. Many kinds of tags are now in use suitable for different kinds of fish, even some with small acoustic transmitters attached.

The other method is to count the number of eggs liberated into a spawning area, as for instance shown on the chart for plaice in the southern North Sea (see p. 212). From a knowledge of the average number of eggs produced by a fish the total number of spawning fish can be estimated. If this figure be multiplied by a factor to allow for males a figure is obtained for the total adult population.

By these two methods a figure of 200 to 300 million has been found for the plaice, which is probably something of the right order. Similarly it was estimated that in 1950 there may have been 813,000 t of pilchards living in the English Channel, or ten thousand million individuals.

The methods described above in detail for studying the plaice and herring are now being used for many other kinds of commercial fish. Regular yearly surveys are made over large areas to plot the distribution of the eggs and planktonic stages of these fish in order to assess the sizes of the stocks. These are reported on and discussed at the annual meetings of the International Council for the Exploration of the Sea in so far as the North Atlantic is concerned, and by other organizations for other oceanic areas.

Although many of the methods employed in fishery research have been mentioned in the above outline of the study of two such fish as the plaice and the herring, there are still others that remain unmentioned. Space will not allow of further expansion on this subject but enough should have been said to impress the reader with the immense amount of labour that is involved in such investigations. We cannot be satisfied with one or two observations only. Day in, day out, information must be gathered from all possible sources; statistics of fish landings, measurements of fish, samples of the bottom, collections of plankton, observations of currents, temperatures and saltness, all are needed. Furthermore we cannot be content with such information gathered over a period of only one year. Observations must be carried through without fail from year to year, almost it seems indefinitely; for until many years are so covered no possible light can be thrown on the causes of the great natural fluctuations of the fish supply, of good or bad survival years.

One of the factors is certainly the availability of the right kind and size of planktonic food organisms for the tiny fish at the critical stage when it has recently hatched from its egg and used up its supply of yolk. It has for instance been shown that in the North Sea the young plaice, when they first start feeding, eat almost exclusively the pelagic tunicate *Oikopleura*. In some years the young may be drifted by water movements into unfavourable areas where they perish.

Research has also shown that it is not sufficient merely to study the conditions of the area with which one is concerned. It is necessary to know the sequence of events in other far lying regions, for all the waters of the oceans bear an intimate relation one with another in their current systems, and an alteration in the trend of a current in one locality may have far reaching effects on the conditions for life in waters many kilometres away. The fisheries of the North Sea are influenced to a large extent by that north-easterly drift of the Gulf Stream which, rounding the north coast of Scotland, turns southwards round the Shetlands into the North Sea itself. It is known that the Gulf Stream shows changes in its strength and in some years water from the North Atlantic Drift may penetrate deeper into the North Sea than in others. On such occasions it may bring with it abundant planktonic life in the mixed oceanic and coastal waters. Such waters are sought after by the herring for food, so that the extent of their penetration into the North Sea will affect the distribution of the feeding shoals. The movements of this Atlantic Drift water will also be determined by conditions in the Arctic. Over the past fifty years there have been marked changes in climate. During a period of amelioration in the 1930's the warmer water extended much farther north and the distribution of such fish as cod and herring became altered. The Bear Isle fishery for cod arose and the herring became abundant off the north coast of Iceland. Now conditions are changing once more and the Arctic waters are cooling. This has resulted in for instance a change in the habits of herring. Between 1962 and 1968 the cooling of the waters north of Iceland caused the Norwegian herring to change the course of their feeding migration away from Iceland towards Jan Mayen, Spitzbergen and Bear Island. More northern fish have appeared again off Plymouth. It therefore becomes obvious that fishery research needs also knowledge of climatic conditions that can be supplied by the meteorologist (see also pp. 62, 200).

So far this chapter has outlined the type of research into fishery problems that is carried out at sea in specially equipped research vessels. There remains another side to fishery research of no less importance. The results of the collections at sea must be worked out under favourable conditions and to this end a marine laboratory becomes a necessity. A laboratory for fishery research must be fully equipped both with staff and accommodation for study in any branch of science. There must be biologists who can deal with the collections of animals and plants made at sea; there must be chemists to study the salinity of the water samples and gain an insight into the movements of the water masses, to study the gases and other dissolved substances in the sea water and note their interrelation with the distribution of animals and plants as shown by the biologists; physical observations of temperature and light penetration must be worked out; and there remain the statistical details on fish populations and catches to be dealt with, for which a special mathematical knowledge is required.

But while such laboratories will be concerned purely with research on the fisheries and the economic applications of science, others are required in which research of a more fundamental character may be carried out. In such laboratories problems, which to the lay mind appear at first sight to have no bearing on fishery research, are studied. It is quite a common occurrence in scientific research that an observation apparently of a very isolated and inconsequent nature may be developed later into an instrument with far-reaching practical results. Opportunity is afforded in these laboratories for workers from universities, from the home and other countries, to study a variety of biological problems, including problems of a medical or other nature.

The investigation into the best methods of preserving fish for food by freezing, and the effects of refrigeration on the actual values of the fish flesh, is a subject that receives considerable attention, and much study of this kind can be carried out in laboratories, results of which are invaluable to the fishing industry.

With the increase in our knowledge of the sensory perceptions of marine animals (see Chapter 10) there has been much attention paid to a study of the behaviour of fish. Research has been done in the laboratory on the vision of fish under different light intensities and on their swimming speeds. At the same time observations are made at sea by means of echo sounders, television and by direct underwater observations by diving or in submersible craft on the behaviour of fish and of their reactions to fishing gear. In this way it is possible to design the most efficient nets for their capture.

Another line which may be pursued in the laboratory is the rearing of fishes. For a long time, for instance, the Port Erin laboratory has run a plaice hatchery in which plaice are reared through the critical period of their lives and then liberated into the sea. Lobsters are reared in the same way in Scotland and Norway where for many years cod have been hatched and liberated in one of the fjords. The study of the methods of rearing marine fishes is now receiving increasing attention in marine laboratories for this is a necessary preliminary to investigations on their genetics. It is one of the aims of fishery research to examine the possibilities of the improvement of fish stocks by breeding selection and perhaps the introduction of hybrids. In this way we may eventually really reach the position of being able to farm the sea. Marine fish farming has in fact now reached the development stage resulting from earlier research by the Ministry of Agriculture, Fisheries and Food at the Lowestoft laboratory and in Great Britain the White Fish Authority has pilot fish farms in operation.

Freshwater fish, such as trout and carp, have for long been cultivated by man. The eggs of these fish are large and yolky, and the fry hatch at a large size and are soon able to fend for themselves and find and capture their food organisms. But the eggs of most marine fish are, as described in Chapter 4, very small and the young on hatching relatively minute. At this early stage their mouths are extremely small and capable of taking only the most minute planktonic organisms. This is a critical stage in their life-history. While it is easy to fertilize their eggs and hatch them under artificial conditions it has always proved extremely difficult to supply sufficient food organisms of the right size to rear large numbers of young. It was only just before the last war that it was found that the newly hatched young of the brine shrimp *Artemia* were sufficiently small to be used as food for some fish. These crustaceans live in vast numbers in salt pans. Their resistant eggs can be collected and stored dry until needed. Thus they can be used as a controlled food supply to be hatched just when required. As a result of this it is now possible to rear such fish as plaice to about the size of postage stamps by the million.

The next step is to find the most suitable method of growing them on to marketable size. Experiments have shown that they can survive in floating cages in the sea in very crowded conditions and given sufficient supplies of food. The problem thus now becomes an economic one. Capital outlay, maintenance costs and price of food all mount up. At present the cost of producing a marketable plaice by this means is higher than that of one caught at sea. But costs of ships, gear and wages are continually mounting and it may well be that in the future plaice produced in fish farms may prove to be the cheaper.

In the meantime experiments are being made with valuable prime fish such as sole and turbot, whose price when caught at sea is high owing to their scarcity and choice eating qualities. It may certainly prove an economic proposition to cultivate such fish. When newly hatched, however, their mouths are smaller than those of the plaice and unable to take young brine shrimps. Search is going on for a suitable food which can be cultivated in sufficient quantities, and rotifers and small worms have been found to be suitable.

Another relevant factor is the time taken to grow to marketable size. Increasing the temperature of the water can markedly increase the rate of growth and experiments are in progress using warm effluent sea water from power stations.

Fish farming also requires a knowledge of the nutrition of fish and research in this direction is being actively pursued.

But the sea is big and the supply of food therefrom would appear to be inexhaustible. It is only natural, therefore, that fishery investigations should have had many critics. It was argued that the expense involved was not worth the gain and that the sea contains resources that can never be used up by man's puny efforts; that even if the fishermen did exhaust one ground they would be forced to move elsewhere for their supply, with the result that the fished-out ground would have time to recover.

Such suggestions appeared perhaps at one time to be sound, but with the ever-increasing world population man is now turning more and more to the sea for his supplies of food. More and better fishing vessels are being built and more efficient means of capture devised. It is known that overfishing can happen and that future stocks can be safeguarded only if the necessary precautions are taken to prevent it. For this purpose there are now international commissions to watch over certain fish such as halibut, tuna and sardines. By careful regulation since 1931, for instance, the annual yield of Pacific halibut has been greatly increased, and would otherwise have probably dropped to under half its present level.

Although oceanographical research is still young there is no doubt whatever that the expenditure has been worth while. We are now in a much better position to understand the general economy of the sea and are no longer entirely in the dark, as we certainly were at the beginning of this century. Investigations, which used to be confined to the areas of the northern fisheries, are now world wide and they are being carried out at an increasing scale. New fisheries are being developed in areas where a foreknowledge of the environmental conditions indicated the likelihood of abundant supplies, and this is a direct outcome of fundamental research.

Many kinds of the well-known food fishes are now indeed being overfished and attention is being turned to the possibilities of making use of oceanic and deep-water species not yet used. Of first importance among these may be the abundant small blue whiting (*Micromesistius poutassou*) which is now beginning to be exploited.

The chief aim at present must be to estimate the limits to which the different fisheries will bear depletion by a continuing watch on the fish populations. It seemed doubtful whether man would ever succeed in actually increasing the productivity of the sea, but even that may now be within the bounds of possibility. It has been shown that the main control of the productivity of a given area lies in the quantity of dissolved nutrient salts present in the upper water layers (p. 175), and an increase in these salts would probably give rise to a greater abundance of fish life. In these days of atomic power it may well be that means for so doing will be found, for there is a vast reservoir of nutrient salts in the deep waters of the oceans to be drawn on. At any rate, the manuring of estuarine waters for improving shellfish cultivation is a practical proposition and experiments on the fertilization of sea lochs have already been started.

It is certain that we are on the threshold of new developments resulting from the accumulating knowledge of the marine environment and we are only beginning to realize the possibilities of exploiting the resources of the sea.

Chapter 16

MARINE EXPLOITATION AND CULTIVATION

The sea is a vast reservoir of raw materials which become increasingly available as technical methods develop, although man has obtained food and a variety of natural products from this source since the earliest time. Undoubtedly the greatest ingenuity in this respect has been, and continues to be, displayed in the Far East although we must here largely confine ourselves to Europe and America. Exploitation of marine resources has primarily been by hunting—of fish, whales and seals as already described in Chapter 14—but there has also been a great deal of straight collecting, aided by the use of simple traps, between tide marks and in shallow waters. Marine invertebrates, notably the molluscan and crustacean 'shellfish', must represent the earliest source of food taken from the sea. Over the last century and rather more, increasing exploitation has outrun natural productivity and throughout the world more and more attention has been paid to cultivation, to systems of mariculture—or aquaculture—similar to agriculture on land. Problems are not so great here as they are with fish, and impressive measures of shellfish cultivation are practised in many parts of the world, some of which are described here.

Other important products come from a diversity of animals and seaweeds, while sea water itself is a source not only of common salt and of magnesium but contains every known element, increasing numbers of which may be obtained as methods of extraction improve. As discovered by the *Challenger* expedition, in places the floor of the deep sea is covered with manganese nodules which one day promise to form a major source of this element. The sea bed is also a source of minerals which are likely to be mined, in the security of underwater chambers, when terrestrial mines become exhausted. Already diamonds are collected from the sea floor. All now know about the rich supplies of natural gas and of oil which lie under the continental shelf off Europe as well as other parts of the world. These matters now affect Western European countries very intimately. It is impossible to deal with these developments here but suitable books for general reading are listed in the Bibliography at the end of this volume.

Shellfish have been consumed by man from time immemorial as indicated by the presence of shell mounds, often of great dimensions, near the dwellings of prehistoric man, and so arranged that they are certainly the work of man because sometimes containing animal bones and rude weapons. On an island off the coast of California a primitive race of people living largely on shellfish are said to have survived until the nineteenth century. Oysters have long been regarded as a delicacy, to such an extent by the Romans that, according to Pliny, they increased the natural supply by cultivating them artificially in Lago Lucrino, and Roman vases from the time of Augustus have been discovered on which are designs showing the method of cultivation.

There are two genera of edible oysters, *Ostrea* and *Crassostrea*. The former are flat, like the British 'native' *O. edulis*; they also incubate the eggs in the gill cavity so that during the summer (when there is no 'r' in the month) they are successively 'white sick' and then 'black sick' due to the appearance of the finely granular developing larvae revealed when the shell is opened. In *Crassostrea*, the lower valve is deeply cupped while egg as well as sperm is liberated into the sea where all development takes place. There is no period when they are inedible. Both types of oyster usually begin life as males, later changing sex; in *Ostrea* there is an alternation in sex throughout life.

In Europe there are two indigenous species: *O. edulis*, which extends from the eastern Mediterranean round the coasts as far as Norway and, along the coasts of Portugal from which it has spread during the past century up the Biscay coast of France, the Portuguese oyster (*Crassostrea angulata*). There are many species elsewhere, the most important being the American *C. virginica* which extends from the Gulf of St. Lawrence to the Gulf of Mexico and the Japanese *C. gigas*, the most virile of all oysters which has been introduced along the Pacific coast of America, where it has in places established itself, and also into Australia and more recently into Europe. Many other species are collected and eaten locally.

Oyster Cultivation

Edible oysters are shallow-water animals with a preference for estuarine conditions. This made them initially easy to collect—they were also enormously abundant—and has greatly facilitated their cultivation which forms an important industry in many parts of the world, especially in France and Japan.

The history of oyster cultivation in Europe is fascinating. As we have already seen, in this, as in so much else, the Romans were the pioneers, and it is probable that the crude methods of cultivation formerly in use in Lago Fusaro, and maintained in the Gulf of Taranto in Italy, have been handed down from the days of the Romans. It was certainly an examination of these by the French Professor Coste which resulted eventually in the development of the great modern French oyster industry. The increasing popularity of oysters among the French in the early nineteenth century led to a more and more intensive fishing of the fine natural oyster beds with which the western coast of France was fringed, and though for many years this seemed to have no effect on the number of oysters collected, yet finally the natural rate of increase was overtaken and stocks began to diminish, to such an extent that, to quote but one example, 70,000,000 oysters were dredged in the Bay of Cancale in 1843, and only about 1,000,000 from the same locality in 1868! All along the coasts of France the oyster beds were depleted in a like fashion, and the government had finally to step in, drastically restricting oyster dredging, pending scientific investigations.

It was the boundless enthusiasm of Professor Coste of the Collège de France, which eventually turned disaster into success. The first experiments consisted merely of attempts to restock the old beds by means of imported oysters. Experiments of this nature carried out near St. Malo in Brittany met with success, and it was realized for the first time, moreover, that young or 'spat' oysters could be collected artificially on fascines or bundles of twigs just as successfully in France as they had been in Italy. To make this clear we must say a little about the life-history of the oyster. The adult oyster is cemented at a very

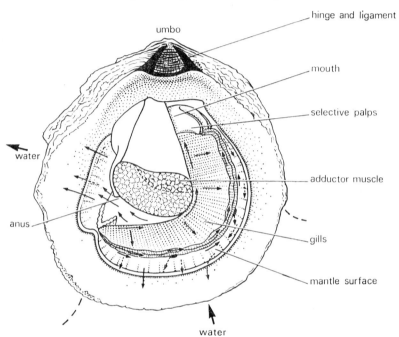

Flat or 'native' oyster, *Ostrea edulis*, viewed lying on left side after removal of right valve and mantle lobe. Major structures labelled; large arrows indicate water currents created by gills; arrows on the gills show the direction of food-collecting currents on their surface (passing to mouth); broken arrows on mantle show surface direction of currents for collection and removal of waste.
(After Yonge)

early age to stones or rocks and can only be moved by currents powerful enough to move the stone to which it is attached. But the young oysters—produced in countless millions, for a single oyster may produce millions of eggs in a single season—are unattached for the first few weeks of their existence and rise near the surface of the sea where they are carried in all directions by tides and currents. At this stage they have a pair of tiny shell valves between which projects the velum or 'sail' covered with cilia, by the rapid movement of which they swim and also collect their food. At the end of their brief period of freedom the little oysters, or such of them as have survived the many hazards, sink to the bottom.

There they must attach themselves to relatively clean, hard objects such as stones and shells; if the bottom is of sand or mud or if the stones are slimy with weed the oysters will die. After some initial crawling by a temporary 'foot', the minute oyster cements itself, left side undermost. It now very quickly metamorphoses into the adult form and begins to grow rapidly, so that soon, although less than 3 mm in diameter, it has all the organs of the adult oyster in miniature. It is now merely a question of time, abundance of food, and immunity from predatory foes such as starfish and crabs, before the oyster becomes fully grown.

The discovery that oyster spat would settle upon artificial collectors like brushwood immediately fired the imagination of Coste. He foresaw that not only would it be possible to restock the depleted beds, but, by collecting spat wholesale, to cultivate oysters on an immense scale. The Emperor Napoleon III became interested and two Imperial oyster parks were established in the shallow Bay of Arcachon, south of Bordeaux. The experiment was a brilliant success; 2,000,000 oysters had been imported and from these immense numbers of spat were obtained on a new type of collector consisting of planks covered with pitch or resin, the latter, with the attached spat, being subsequently broken away from the woodwork. Later years, however, were not so successful but a great step forward was made when half-cylinder roofing tiles were adopted as 'collectors'. These were later coated with a mixture of lime and sand which could be flaked off without damaging the delicate spat. When so detached these were placed in 'ambulances' consisting of shallow boxes on short legs and enclosed with fine wire netting where they remained until they were laid out on the parks. But with high labour costs these ambulances have now been abandoned.

The Bay of Arcachon became the greatest centre of oyster culture, over 500 million being marketed annually. The bay is almost land-locked with only a small opening. At high tide the sea covers some 14,800 ha, more than half of which is exposed at spring tides and is covered with oyster parks such as those shown in Plate 46. Each is surrounded by a palisade of branches which serve to protect the oysters from the attacks of marauding fish, largely rays, and which provide the only indication of the presence of these areas of intense cultivation when the tide is full. Cultivation consists of raking the oysters so that they are raised above the continually falling silt (because these beds are artificial) and removing the ubiquitous crabs and starfish, the greatest enemies of young oysters. In early summer the collectors (Plate 46) are put out in preparation for the spat which will settle after the water becomes warm enough for the oysters to spawn (when the temperature rises above about 15°C). The tiles are scraped and relimed and then arranged, concave side downward in crate-like wooden cases. Great numbers of these collectors are put out just below low-water mark, where the tiny swimming oysters will be washed against them every time the tide retreats, and if there is a good 'spat-fall' the tiles soon become covered with thousands of transparent specks, each representing a potential marketable oyster.

The collectors remain in the water until the beginning of the following year, care being taken to keep them clean and free from encrusting plants and animals. As soon as the weather is good, the tiles are brought ashore, so many at a time, to a shed where 'détroquage', as the process of separating young oysters from

Q

the collecting tiles is called, is carried out. For the time being the oysters are placed on wooden trays in a storage tank, but as soon as possible they are picked

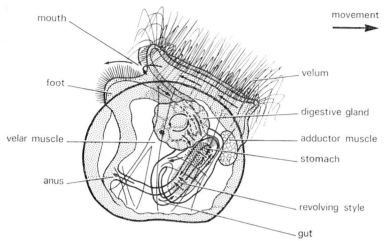

Veliger larva of oyster showing velum protruded with large cilia with which it swims. Major organs labelled. Arrows indicate course of feeding currents round base of velum and movements in gut. (After Yonge)

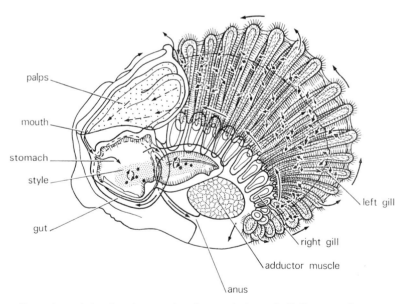

Recently settled or 'spat' oyster showing newly formed gill filaments, those on the left forming before those on the right, and very large palps at this stage. Arrows indicate direction of ciliary currents on gills, palps and throughout the gut. (After Yonge)

out, cleaned and separated into different sizes and laid on the surface of the parks.

Oyster cultivation is carried on in many other regions along the west coast of France, notably off the Ile d'Oléron south of Rochfort and along the south coast of Brittany. The fattening of oysters forms a separate industry in certain parts, the most important of which is in the district round the mouth of the River Seudre, near Marennes and La Tremblade. Here the land is flat and clayey and in it are dug series of shallow basins of 'claires', into which sea water enters during the period of spring tides. During the intervening periods the water remains stagnant and in summer becomes very hot and, by evaporation, very salty, conditions which result in a tremendous increase in the microscopic plant life, notably diatoms, which forms the ideal food for the oysters that are laid on the bottom of these claires. These grow and fatten at an exceptional speed and soon reach a marketable size, when they are taken out of the claires and placed in stone tanks, known as 'bassins de dégorgement', containing fresh, clean sea water in which they are left for several days in order that they may become thoroughly cleansed internally.

It is in the claires at Marennes that the well-known green oysters are produced. These obtain their colour, which has become known as Marennin, from a particular diatom named *Navicula fusiformis* var. *ostrearia*. Although there is little discernible difference in taste, these green oysters are considered a delicacy in many places and command a high price.

Originally the native oyster was exclusively cultivated in France, but became almost completely ousted in the more southern beds by the Portuguese oyster of which the vast majority of oysters reared at Arcachon or fattened at Marennes now consist. The Portuguese oyster obtained its footing in France in a peculiar manner. A cargo of these oysters, considered to have gone bad, was thrown overboard near the mouth of the Gironde, but some at least recovered, for a number of years later a large and flourishing bed of Portuguese oysters was discovered on the spot where they had been thrown overboard. In the more northern beds, mainly on the north and south coasts of Brittany, where the sea temperature is too low for the Portuguese oyster to spawn, the flat *O. edulis* (which, incidentally, is more highly prized and fetches a better price) is still cultivated. It suffers from being less hardy, while it cannot be cultivated between tide marks. It was also considered more prone to the protozoan and fungal diseases which have devastated oyster stocks, probably in Britain after the first world war and certainly subsequently in Holland and along the Atlantic coast of the United States. However, the Portuguese has now succumbed to a similar disease, and the livelihood of thousands of French growers is threatened. At the time of writing (1974) attempts are being made to restock the Biscay beds with the supremely hardy Japanese *Crassostrea gigas*.

Although Great Britain was once ringed with rich oyster reefs, these were over-fished and nothing comparable with the French system of cultivation was established despite various attempts. Stocks have continued to decline so that today they are only a few per cent of what they were a century ago. One major cause of decline has been the spread of pests, largely brought in with imports of

American oysters, notably the oyster drill, *Ocenebra* (p. 138) and the slipper limpet, *Crepidula*, which forms chains, competing with oysters for planktonic food and literally smothering them with their great numbers. The principal beds are in Essex and Cornwall; everywhere the oysters live in shallow water but are seldom uncovered in any number even at the lowest spring tides. They are usually collected by small rowing or sailing boats with the aid of iron dredges which are dragged along the bottom, the oysters being collected in the dredge bag which is usually made of steel links. Spat collectors are seldom used, but during the summer it is customary to throw great quantities of clean shell, known as cultch, over the beds so as to provide good settling surfaces for the spat. However, it is perfectly feasible to cultivate oysters in Britain. To this end the Ministry of Agriculture, Fisheries and Food in parallel with work conducted elsewhere, notably in the United States and Japan, has conducted experiments at Conway in North Wales where the White Fish Authority established a pilot hatchery. Here oysters, both *O. edulis* and the Japanese *C. gigas*, are induced to spawn at all times of the year by appropriate rise in temperature and the larvae are then fed on suitable cultures of minute green flagellates and of diatoms which experiments had proved to be adequate food. After settlement the spat are now immediately detached and then properly fed until large enough to be placed, suitably protected, in the sea. Several oyster hatcheries based on these methods are now in use.

Oysters were extensively cultivated in Holland, around the mouth of the Schelde which is now being enclosed by a dam. In Norway they are cultivated in pools cut off behind glacial moraines at the head of fjords and which warm up greatly in the summer, the microscopic plant life increasing as it does in the claires. The oysters are suspended in midwater, the surface water being too fresh and the bottom water devoid of oxygen.

The Atlantic coast of North America was originally fringed with dense beds of oysters on which a great industry was based, although all well below sea level, the oysters being collected by dredges or by means of long wooden 'tongs'. However, of recent years populations have been drastically reduced by the inroads of pests and especially of protozoan and fungal diseases, some of them still very obscure. There has recently been a marked increase of oysters in Long Island Sound where a highly successful hatchery has also been established. Very great numbers of these American oysters (*C. virginica*) are still marketed, usually after 'shucking', namely removal from the shell. On the Pacific coast the indigenous oyster is the small Olympia flat oyster (*Ostrea lurida*,) but this has been almost completely ousted by the import, and in places, including Canada, successful establishment, of the Japanese *C. gigas* which all too easily grows too large for marketing.

This oyster, although only one of several species which occur there, is cultivated on an enormous scale in Japan (Plate 48). This practice began perhaps two centuries ago on palisades of bamboo on which the young oysters settled and from which they were eventually harvested. Today oysters may be cultivated on stakes driven into the muddy bottom of shallow inlets or—and this is now the commonest method—they are suspended in long strings from rafts maintained on the surface by large concrete floats. This method has the major advantage of keeping the oysters clear of the bottom and so away from contact with pests which cannot swim, such as starfish and drills. In either case oyster spat is collected on large scallop shells which have a hole drilled in the middle and are

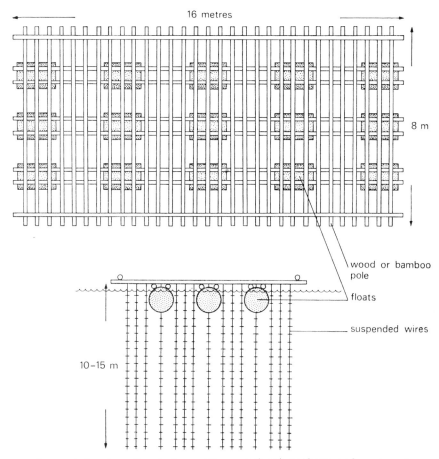

Suspended oyster culture in Japan. Diagrams showing: above, surface view of raft of crossed poles buoyed up by concrete floats; below, side view of raft showing suspended wires to which large scallop shells bearing attached oysters are secured. (After Fujiya)

threaded on to wires with 'spacers' in between. The oysters tend to grow irregularly, but since all are removed from the shell before sale this does not matter. Oyster 'seed' which is exported, largely to the United States and Australia—but now also to France—is specially hardened and is shipped in trays which are hosed down with sea water during the voyage. This Japanese oyster is in a fair way to dominating the world; it is certainly easier to cultivate and to grow in Britain than is the 'native', but it needs a considerably higher temperature for natural spawning.

Other Molluscs

After the oyster, the common mussel (*Mytilus edulis*) is the most important molluscan shellfish used as food in Europe. It is so common as to need no description. It is very abundant in sheltered parts of the coast, or in the mouths of rivers along the coasts of Britain. Natural mussel beds containing animals of all sizes may be 15 cm deep. The shell is not itself attached, but the animal discharges a series of fine, viscid byssal threads which are directed by way of a groove on the thin, extensible foot on to a convenient surface to which they adhere and quickly harden to form an extremely strong attachment so that it takes a violent storm to break up a mussel bed. So rich are these beds that it is seldom necessary to cultivate mussels, although small, stunted mussels from overcrowded areas speedily grow to large marketable size when transplanted to new areas. Many mussel beds are exposed at low water and the mussels collected by hand, but others are always underwater so that they have to be fished from boats by means of dredges.

Mussels are cultivated on the west coast of France in the shallow, muddy bay called Anse de L'Aiguillon (Plate 46). The system of cultivation employed dates back over seven hundred years to the chance which wrecked an Irishman called Walton, travelling on a small vessel containing sheep, in this region in 1235. He and some of the sheep—from which it is said certain valuable modern flocks are the descendants—were the only survivors and, in order to obtain a living, he tried to snare sea birds by means of rough grass nets fastened to stakes on the muddy shores. It is doubtful whether he caught many birds, but quite incidentally he discovered a more reliable food supply, for young mussels speedily covered his nets and, because they were raised above the mud in which they would otherwise have been smothered, they quickly grew to edible size. From this strange beginning was elaborated the modern method of mussel cultivation on stakes interwoven with twigs, which now extend in parallel rows for hundreds of kilometres giving employment to the inhabitants of several villages. This is known as the bouchot system, and the boucholeurs attend to their mussels at low tide, going out to the bouchots in small flat-bottomed boats called acons which are usually pushed along with the help of one foot encased in a large sea boot.

Since the last war a major industry based on mussel cultivation has grown up in the deep fjord-like 'rias' in Galicia in north-west Spain. Particularly in the Ria Vigo and Ria Arosa the surface of the sheltered water is covered by rows of moored 'rafts' from the under surface of which hang down up to 800 ropes originally of esparto grass but now of nylon. Initially pontoons were used, but these have now been replaced by catamarans with surrounding staging or by rafts upheld by floats. There are now some 1800 of these and with some 50 kg of mussels on each rope this is massive production. The mussels which settle are twice thinned out and transplanted to new ropes before the mussels attain marketable size. This takes place in about eighteen months; the water is warm and there is never any lack of suitable planktonic food so that growth is uninterrupted. An industry now worth many millions of pounds annually has been built up, the majority of the mussels being canned. This is marine cultivation at its best; the mussels are feeding directly on the primary marine production of plant plankton and it was interesting to observe that the pontoons* all incline somewhat towards the side facing the open sea where the plankton is the richer, and so the mussels on that side grew a little faster. Attempts are being made at developing a similar culture in the lochs of western Scotland.

Large bivalves generally known as clams are extensively collected in North America where they are considered a great delicacy. Two species are chiefly eaten, one known as the soft shell clam (*Mya arenaria*) (p. 27), and the other known as the hard shell clam (*Mercenaria mercenaria*). There is now a fishery for this at Southampton thanks to a warm effluent from a power station. The soft shell clam burrows in mud or soft sand from which it has to be dug, and in a favourable area the population is extremely dense. In various parts of the north coast of America clam culture is practised, young or seed clams being planted in favourable localities, from which the harvest is reaped in due course. In this way the soft shell clam, originally an Atlantic animal, has been introduced to the Californian coast where it has found ideal conditions and has spread over great areas round San Francisco Bay. The hard shell clam never appears to form such dense colonies and it has become comparatively scarce owing to the reckless way in which it has been gathered, and it is not yet cultivated to the same extent as the other clam. A wide variety of most appetizing species such as the razor clam (*Siliqua*), the pismo clam (*Tivela*), the butter clam (*Saxidomus*), the horse clam (*Schizothaerus*) and even the great geoduck—largest of all deep-burrowing bivalves)—(*Panope*), are enthusiastically dug in beaches and in muddy bays along the Pacific coast of North America.

Of all molluscan shellfish the scallop, with its rounded margins and radiating grooves, is the most beautiful, and the shell has been used for ornamentation and as a design since the beginnings of civilization. Even the body within the shell, which in the oyster is hardly beautiful, is here highly attractive with its areas of red, white and orange, marking the position of the reproductive organs, the muscle and the gills respectively. Scallops are highly prized as food, two species being marketed in the British Isles, and many others in different parts of the world. Of the two former, one, the common scallop (*Pecten maximus*), of considerable size, 10 cm or more in diameter, is rightly considered a luxury, while the other (*Chlamys opercularis*), about a quarter that size, is often captured in large numbers off the south coast where it is known locally as the queen and finds a ready sale, being eaten raw like the oyster or else boiled. Scallops are no less popular in the United States, but there only the adductor muscle is eaten.

The edible cockle (*Cerastoderma edule*) has a rounded white shell about 2·5 cm long covered with small protuberances which enable it to grip the sand. It is extremely common between tide-marks in sheltered stretches of sand, the hind end of the shell with the short siphons just flush with the surface. The foot

* The catamarans may not be so obviously affected.

is a large and powerful organ with which cockles can move very actively on the surface as well as just below. If there are adequate supplies of planktonic food in the water, the sand may be literally packed with cockles lying 'cheek by jowl'. A bed of some 128 ha in South Wales had, when surveyed, an estimated population of 462 million cockles! After a brief larval period when they are carried widely by tides and currents, the young cockles settle, tending to concentrate in areas of soft, oozy mud which provide ideal nurseries for these 'seed' cockles which form a compact layer just below the surface. As they grow, the boundary of the nursery extends, on all sides over the surface of the sand but in the centre, where there is no possibility of further extension, the surplus population is forced to the surface whence it is largely carried away by currents, part to assist in the restocking of the sandy beds, and part to destruction. Cockles have the common enemies of all such shellfish; they are eaten by starfish, bored into by dog-whelks, and eaten by oyster-catchers and gulls, and probably also by flat-fish which swim over the beds when the tide is up. Owing to the exposure

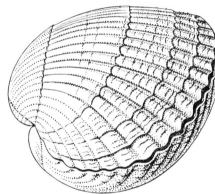

Cockle, *Cerastoderma* (*Cardium*) *edule*, showing annual and disturbance rings on the very rotund valves

of the sandy beds, frost in winter or great heat in summer are both fatal to the cockle. Storms and currents also do great damage, the former by throwing the cockles above high-water mark and the latter by washing them from favourable to unfavourable localities.

The most important cockle fisheries in Great Britain are in the Wash, the Thames estuary, Carmarthen Bay and Morecambe Bay. The cockles may be scraped up by hand, or may be dug, but more often short-handled rakes with large teeth or scrapers consisting of sickle-shaped pieces of iron are used. In the Thames estuary a method of suction dredging for cockles in shallow water has recently been successfully developed. Cockles are seldom sent alive to the market, but are usually boiled, often in iron vessels formerly over extemporized fireplaces on the sea shore but now increasingly by more efficient methods.

The age of a cockle may be judged from its shell. This is being added to continually but more rapidly in the summer, when food is more abundant and temperature higher; then the individual additions are spaced more widely. During the winter these are much closer, forming well marked 'winter rings' as shown in the figure. However, if growth has been seriously disturbed—if cockles have been dislodged by storms or thrown too far up the beach—'disturbance rings', not distinguishable from winter rings, are produced. Since young cockles can extend the edge of their shells at the rate of 1 to 1·5 mm per week it is easy to see how a few days' disturbance may interfere with the even production of the shell. Mussels and other bivalves show similar disturbance rings but not always winter rings. Oysters increase their shells by sudden bursts of activity, depositing a broad but very thin layer of new shell round the edge at a single operation. This thin layer is called a shoot and quickly thickens and hardens. The number of shoots per annum varies, but in this country appears to be between two and four, each of which in an oyster of 5 to 8 cm in depth may mean an increase of about 8 mm all round the edge. The variable number of these shoots makes it very difficult to judge the age of an oyster unless the local conditions are very well known.

Hitherto all the molluscs we have considered have been bivalves, but there are also a number of univalve molluscs, marine members of the snail family, which are used as food although only to a minor extent in Great Britain. The most important is the common periwinkle (*Littorina littorea*) which is extremely common on the middle and lower regions of rocky shores, where it is gathered at low tide, being boiled before it is eaten. The common whelk (*Buccinum undatum*) is much larger and far less common on the shores, but is taken by dredging or in pots. It has a strong white or yellowish shell up to 10 cm long and 6·5 cm broad. The contained animal is correspondingly large but the flesh is close and tough. Nevertheless, after boiling or pickling, it is eaten in large numbers. The common limpet (*Patella vulgata*) is also eaten but it is very tough.

Overseas it is very different. One of the most prized delicacies off many Pacific coasts, in California, Japan and New Zealand, are the large abalones (species of *Haliotis*). These are really great flattened limpets, adhering most tightly to rocks, with a curved series of openings through which water leaves the gill chamber. The larger species are up to 20 cm long and the massive foot, after suitable hammering to break up the muscles, makes an excellent dish when fried. A smaller species (*Haliotis tuberculata*) occurs in the Mediterranean, coming as far north as the Channel Islands where it is known as the ormer and is carefully protected.

Everywhere in the Gulf of Mexico and the Caribbean are rich, almost it would seem inexhaustible, populations of the massive conch (*Strombus gigas*), the thick red shell weighing up to 2·7 kg. This is marketed both fresh and after sun-drying and makes excellent chowder. Eggs are laid in masses, containing up to 750,000 individuals and development is direct with the young emerging as miniatures of the adult, to the full size of which they may grow in four years. They feed on delicate algal growths and should natural stocks fail would be well worth cultivating. Other large snails are eaten in many parts of the world, including big top shells (*Batillus*) in Japan.

We should not leave the subject of food from molluscs without mentioning squid, and also cuttlefish and octopods, all of which are eaten in enormous quantities, especially by Mediterranean and oriental people (p. 192).

Molluscan shellfish are highly nutritious, storing large quantities of glycogen ('animal starch') as well as fat and protein and so forming a complete food in themselves. An oyster has been compared in food value to a glass of milk, while a report from the Ministry of Agriculture, Fisheries and Food stated, 'It has been calculated that an acre (0·4 ha) of the best mussel ground will produce annually 40,000 lb (18,144 kg) of mussels, equivalent to 10,000 lb (4536 kg) of mussel meat with a 'fuel' value of 3,000,000 calories . . . and this at the cost of practically no capital expenditure and only such labour as is involved in transplantation to prevent overcrowding, and to secure the best conditions for growth and fattening. No known system of cultivation on agricultural land can produce corresponding values in the form of animal food. The average yield in beef of an acre (0·4 ha) of average pasture land is reckoned to be 100 lb (45 kg), equivalent to 120,000 calories. . . . The yield of rich fattening pasture may be as high as 190 lb (86 kg), equivalent to 480,000 calories. . . .'

Purification

Unfortunately, there is one great drawback to the consumption of shellfish, the risk of bacterial infection. Oysters and mussel beds are often situated in estuaries many of which are badly contaminated by sewage brought down from towns farther upstream. Any bacteria are liable to be taken in by the shellfish in the course of their normal feeding activities. Though they are themselves unaffected, they may in this manner become a medium for the dissemination of typhoid and other pathogenic bacteria. Rigorous regulations forbid the sale of polluted shellfish but the risk of disease has been an important and, in the past, not unjustified, reason for the undoubted prejudice against the use of shellfish as a food.

The best preventive measure would be to forbid the discharge of unpurified sewage in the neighbourhood of shellfish beds, but unfortunately this is still impracticable. The only alternative is to purify the shellfish and this has been done with striking success at Conway, where the extensive mussel beds had led to the development of a considerable industry, the collected mussels being sold to the large Midland towns. After an adverse bacteriological report the sale of these mussels was prohibited in 1912 and many men were thrown out of employment with the result that investigations were started in the hope of developing a suitable method of purification. Large concrete tanks were constructed and, after lengthy experimentation, a simple but eminently successful process of purification was established.

The mussels are spread two deep on wooden grids in shallow tanks and thoroughly hosed to clean the outside of the shells. Sea water, sterilized by the addition of chloride of lime, any free chloride being converted into common salt by the action of sodium thiosulphate, is then run in by gravity from storage tanks, and the mussels are left in the water for one day. They quickly open their shells and discharge the contents of the gut and of the gill chamber, including any contained bacteria, while at the same time they take in sterilized water. This water is then run off and the tank filled with fresh sterilized water for a similar period to make sure that the mussels are thoroughly cleansed internally. Again the water is run off and, last of all, the shells are sterilized with chlorinated water, the cleansed mussels being then packed in sterilized bags. This system of cleansing was begun on a commercial scale in 1916, and has proved a complete success; the flavour of the mussels is not impaired while there have been no further complaints of pollution, and the mussel fishery has regained its original importance. As expenses have to be met by a fixed charge for each bag of mussels cleansed, only the larger mussel fisheries can be treated in this way owing to the expense of erecting the cleansing tanks, and consequently many of the smaller fisheries which have been condemned cannot be re-opened.

Oysters do not appear to take in bacteria quite so readily as do mussels but they were initially more difficult to purify when using similar methods, largely because the temperature had to be higher. But of recent years excellent new methods of cleansing have been devised using ultra-violet light. Some 5000 flat or Portuguese oysters can be cleansed in a day in a tank no larger than 6 m long by 3 m wide.

Crustaceans

In total bulk and value the global produce from crustacean fisheries is now much greater than from molluscs. This is due to enormous developments, particularly since the last war, both in far northern and in tropical waters. In the north great new stocks of crabs have been exploited while the produce from shrimp fisheries in the tropics has become a major export from a number of developing countries as well as a major fishery in the United States.

The largest edible crustaceans are the 'king crabs' of the North Pacific; not true crabs but related to the hermit crabs with an asymmetrical under-tucked abdomen. *Paralithodes camtschatica* is an impressive animal with a massive rugose shell up to 23 cm in diameter and armed with formidable claws. The original populations were enormous and the easier to catch because they collected in great masses in the spring for moulting and reproduction. Effectively starting after the last war, the fishery has been exploited by the United States (largely from Alaska), Japan and the U.S.S.R. and reached a total production of over 160 million pounds (73 million kg) in 1966 but is now (1973) declining probably as a result of overfishing. The meat is marketed either frozen or canned.

In more temperate waters the largest, and most valuable, of edible crustaceans are the European (Plate 41) and American lobsters (*Homarus vulgaris* and *H. americanus*). The latter is somewhat larger but otherwise the two are so much alike that they may well be the same species. Both live in comparatively shallow water and are especially common where the bottom is rocky and uneven; they find protection, especially when moulting, in crevices. They are omnivorous feeders. They spawn in the summer, the fertilized eggs being retained on the underside of the abdomen (tail) for some nine months, the young emerging in a late larval stage and soon moving from the surface waters to the sea bottom. This lobster occurs along the coasts of Europe from the Mediterranean to Norway but is really a northern animal and the most important lobster fisheries occur off the coasts of Norway, Scotland, England, Ireland, and Heligoland, the usual methods of capture being by lobster pots or creels. Some of these are of wicker-work, rounded with a flat bottom and a funnel-shaped opening on the top—rather like a safety inkwell—so that the lobsters can easily enter but usually fail to find their way out; others are in the form of a half-cylinder with a framework of wood and netting. These traps are usually fastened together with rope having

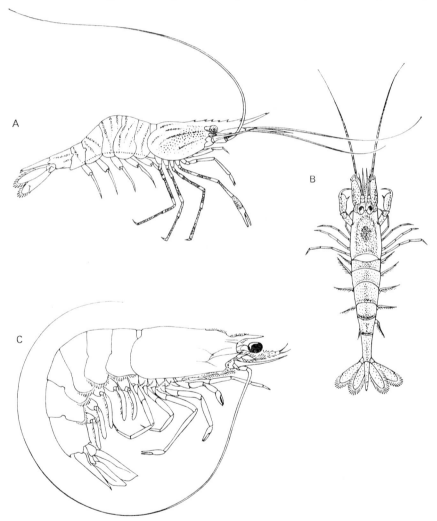

Edible crustaceans : A. Common prawn, *Palaemon serratus* ; B. Common brown shrimp, *Crangon crangon* ; C. Penaeid prawn or 'shrimp', *Penaeus japonicus*, one of many species caught in great numbers in warm and tropical seas.
(A, B × 1, C × ½)

cork buoys at the ends, and before they are placed in position are baited with pieces of stale meat. These lobsters may attain a weight of 4·5 kg, and animals of 6·4 kg have been caught, but the American lobster may attain a weight of 18 kg. It is found on the Atlantic coast of North America from Labrador to New England, where its collection and canning forms an important industry. Recently a productive fishery has developed along the edge of the continental shelf about 160 km from the coast. In many regions lobsters have been caught in such numbers as seriously to affect the fisheries, and protective measures of various kinds have been tried; in some places it is illegal to take from the sea lobsters which are below a certain size or females which are carrying spawn, or are 'in berry'.

In the Hebrides lobsters are collected and stored in large boxes, with holes to allow free access to water, which are moored just awash in the sea. On the Island of Luing, about 23 km from Oban a pond has been made by erecting granite dams at either end of a narrow channel between two islands; a constant flow of sea water is ensured by means of gratings in the dams, while sluices enable the pond to be practically emptied at low tide. One hundred thousand lobsters can be stored here. Similar storage ponds for lobsters have been established for many years along the coasts of Brittany. In Scotland packing is on Kerrera Island 3 km from Oban where there is a sheltered bay in which the lobsters are confined in large wooden rafts each capable of holding about a thousand, until needed for packing. In this process, bracken or fern, which is better than the seaweed or sawdust originally used, is employed, the boxes being lined with paper to protect the lobsters from extremes of temperature. Since it is the presence of water in the gill chamber which is responsible for the animals remaining alive out of water, care has to be taken not to squeeze this region during packing.

In the south-west of England the place of the lobster is to some extent taken by the handsome spiny or rock lobster (*Palinurus elephas*), which differs from the lobster in its larger size, its brown, beautifully sculptured shell and lack of large claws (Plate 41). Seldom eaten in England, it is regarded as a great delicacy in France, where it is known as 'langouste'. There are many species of spiny lobsters in warm and tropical seas, often brilliantly coloured. They are caught in great numbers in different parts of the world such as South Africa, New Zealand, Tristan da Cunha and India and often exported, frozen or canned, usually to the United States.

The Norway lobster or 'Dublin prawn' (*Nephrops norvegicus*), smaller than the true lobster, orange coloured with red and white markings and long slender claws (Plate 41), is common off the more northerly coasts of Europe and also in the northern Adriatic. It inhabits greater depths, from 60 to over 100 m, living on mud into which it burrows. At one time only occasionally caught by trawlers returning to port, this animal is now the main object of a large fishery the product of which is sold as 'scampi', the name used by the Italians who catch it in the Adriatic. In consequence of this intense fishing, the size of the animals caught has diminished very considerably. However, the females appear to burrow while they are carrying eggs and this may constitute some degree of protection against overfishing.

In the British Isles the edible crab (*Cancer pagurus*) comes second in importance to the lobster. It lives in shallow water along rocky coasts under much the
same conditions as the lobster, and is caught in pots in the same way. Fine
specimens may weigh as much as 5·4 kg and be over 25 cm broad. No other crab
is eaten to any extent in the British Isles, although the shore crab used to be sold
in great numbers and is still eaten on the shores of the Mediterranean and on the
Atlantic coasts of France, where the large spider crab (*Maia squinado*) is also
eaten in spite of its extremely thick shell. The large *Cancer magister* is the most
important edible species along the Pacific coast of north America, while the blue
crab (*Callinectes sapidus*), one of the swimming crabs, takes its place along the
Atlantic coasts, where a variety of other crabs is also eaten, some being kept
until they moult when as 'soft-shelled crabs' they are cooked intact. Great
numbers of crabs are eaten the world over; in the tropics these are often swimming crabs which inhabit mangrove swamps into the mud of which they burrow.

Shrimps and prawns are names given somewhat indiscriminately to the
smaller relatives of the lobsters. In Great Britain shrimps are those that live on
and burrow into sand, while prawns live on rocky bottoms, but this distinction is
not universally made. There are two types of these animals. There are carideans,
to which all northern species belong, which, like the lobsters, carry the eggs
attached to the swimming appendages under the abdomen, and the largely tropical
penaeids which grow much larger and where the eggs, fertilized as they issue
from the female by sperm previously applied by the male, are liberated freely into
the sea.

The largest British prawn (*Palaemon serratus*) (p. 24) may exceed 10 cm in
length and has a forward-projecting and obviously serrated rostrum. *Pandalus
montagui*, a handsome animal caught mainly in the Wash, is distinguished by its
pink colour while another species, *P. borealis*, is imported from Norway. This
comes from deeper water in the fjords and the important fishery for it originated
in the work of the Norwegian Fishery Department which revealed its hitherto
unknown presence in deep water and showed how it could be caught commercially in special trawls. Our common brown shrimp (*Crangon crangon*) (p. 26)
is found all around the coast on sand where it is often difficult to see owing to its
greyish-brown colour and habit of burrowing with only the antennae projecting.
Shrimps are caught commercially in large nets attached to long handles which
are pushed over the sand in shallow water. In some regions a large type of shrimp-
trawl is drawn along behind a horse and cart.

The penaeid 'shrimps' are the object of enormous fisheries worth many
millions of pounds annually. These take place in many tropical or semi-tropical
seas, off the Gulf coast of the United States, off both Atlantic and Pacific coasts
of Mexico, in south-east Asia, off southern India and Pakistan, in the Persian
Gulf and elsewhere. All species—and there are many of them—are considerably
larger than cold water prawns reaching up to 20 or 25 cm in length so that two
individuals form an adequate meal. They have a complex life-history made up of
many moulting stages and also involving considerable movements. They spawn,
releasing the fertilized eggs freely into the sea, well off-shore and the young then
move towards the land to feed and rapidly grow to maturity in the shallow and

usually productively rich waters of inshore lagoons where they are often caught in nets or traps when they start to move out for the off-shore spawning migration. The adults are also caught in vast numbers by trawling in moderate depths. One species, *Penaeus japonicus*, is cultivated in Japan. Mature males and females are caught and the fertilized eggs they produce are maintained in suitable aerated containers. The early stages in development are fed with cultures of diatoms, the later ones with animal plankton. When the adult form is attained the animals are put into long concrete troughs with a sandy bottom and in a continuously running and powerful current of water. They are fed on trash fish and on crushed-up molluscs and within a total period of around six months are some 15 cm long and ready for the market. The cost of the process is reflected in a correspondingly high price for the shrimps. In Great Britain prawn culture is developing. It is not difficult to rear caridean prawns; the main danger is cannibalism. Imported penaeids grow much quicker but as yet there is no means of obtaining eggs; all have to be imported in the young stages.

All Crustacea are edible and this covers barnacles. Two very large acorn barnacles, one (*Balanus psittacus*) which occurs on the Chilean coast and the other (*B. nubilis*) along British Columbia, are extensively eaten while a stalked species (*Pollicipes cornucopia*) is eaten in France and Spain.

Pearls

Although crustacean shellfish are valuable only as food, this is not true of the molluscs. They are responsible for the formation of the most valuable of marine products, namely pearls. Although the modern appreciation of pearls probably originated with the Romans, they are mentioned in the literature of most of the early civilizations so that to seek the beginning of the use of pearls we should probably have to go back before the birth of history to the days of primitive men who lived on shellfish and discovered in them round beads of great lustre and beauty.

Many bivalve molluscs form pearls but only in a few are they of the necessary size and quality. The true pearl oysters are species of the genera *Pinctada* and, to a lesser extent, *Pteria*. Although rounded like edible oysters they are much closer to our common mussel being similarly attached by byssus threads and not cemented. The inner shell layer in these bivalves is nacreous, a crystalline structure which breaks up light into its component wavelengths of different colours so giving the 'mother-of-pearl' effect. Pearls are composed of concentric layers of this substance which accounts for their lustre. Pearl oysters are widely distributed throughout the tropical Pacific and Indian Oceans. Valuable pearls also occur in some freshwater bivalves, but they do not concern us here.

The formation of pearls has always given rise to speculation. It was the belief of the ancient Hindus that dewdrops which fell within when the oyster opened its shell were later converted into pearls by the rays of the sun, while another early theory attributed their formation to the action of lightning. But it has long been known that pearls are really due to abnormal stimulation of the tissues of the oyster. Blister pearls, which are fastened to the inside of the shell, and are often sawn off and used in cheap jewellery, are probably a result of sand grains which have worked their way in between the shell and the soft mantle which lies against it and produces it, and pearls, formed *inside* the mantle tissue and so unconnected with the shell, may be produced in this manner. Parasites can certainly cause the formation of pearls, a view which was first advanced in 1554 and has been proved by the discovery of mummified remains of different parasites in the centre or nucleus of many pearls. The parasite which most frequently causes pearl formation in the Ceylon pearl oyster is the early stage of a tapeworm which becomes adult in a large ray which habitually feeds on these oysters so infecting itself with the tapeworm. After becoming mature, the tapeworms lay eggs which pass into the sea where they may find a temporary home in

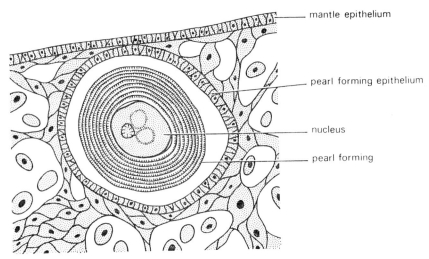

mantle epithelium

pearl forming epithelium

nucleus

pearl forming

Diagram showing the manner of pearl formation

an oyster. A temporary home if the oyster is swallowed by the ray shortly after, but a grave if the oyster should be provoked to protect itself from the irritating parasite by coating it with nacre. It may be that pearls are also formed round tiny granules of waste matter produced by the oysters themselves.

The process of pearl formation is perfectly straightforward. The nucleus, of whatever irritating substance that may happen to be, becomes enclosed within a tiny chamber lined by an inpushing of the superficial mantle tissues which form mother-of-pearl. This is applied in layers around the nucleus so that a round bead is gradually formed, the lustre of which is due to the concentric layers of the crystalline nacre.

Today pearls come largely from Japan and from oysters which have been stimulated to produce them, but formerly this was a haphazard business with divers making indiscriminate collections of pearl oysters. There were many such fisheries, among the most famous that in Ceylon. There the government controlled the oyster beds permitting a fishery only when stocks were adequate which was not frequent; there were only six between 1891 and 1928. Indeed the words of an early writer on the subject named James Steuart, appear only too true: 'It is only when, in the infinite wisdom of the Creator of all things, the oyster brood descends upon the banks suited to nourish and support it—that it comes within our limited power to watch its advancing age, and to fish up the respective deposits in succession as they approach the proper age; not letting them rest on the banks until they die off, and the pearl is lost; and carefully abstaining from disturbing those that are too young to contain it.'

When a fishery was to be held, this fact was advertised in many different languages throughout the newspapers of the East. On the announced date a motley collection of divers, pearl buyers, speculators, moneylenders, shopkeepers and all manner of others, collected at the temporary town of Marichchikkaddi, the headquarters of the fishery. This lasted for anything up to three months and was carefully controlled by the government. The boats with the divers went out every morning and had to return at the sound of a gun about midday, the oysters being landed and the catch immediately divided into three equal piles, two of which were taken by the government and later sold by auction, and the third by the divers. The separation of the pearls from the oysters was a slow and disgusting business, the oysters being left to rot. After a week the largest pearls were picked out by hand, the remaining filth with the smaller pearls washed out of the shells and, after a series of further washings, all the pearls gradually collected. Some such procedure was followed in many other regions, such as the Persian Gulf and around Panama, where such fisheries formerly flourished.

Production of pearls by methods introduced by Mikimoto about the turn of the century is now a major industry, worth very many millions of pounds annually, in Japan. Pearl oysters, here *Pinctada fucata*, are collected both on 'collectors' on which the spat settle or directly by divers who are all women. When a suitable size, the valves are held open while, by a deft operation, a calcareous bead enclosed in a bag of mantle tissue, outer surface innermost, is inserted deep in the tissues. The operated oysters are kept for some years in trays suspended from rafts resembling those used in the cultivation of edible oysters. In the great majority of cases the graft 'takes' although the quality of the subsequent pearl varies greatly and these are rigorously graded. These pearls differ from natural ones only in the nature of the nucleus. Of recent years this culture has extended to north Australia where the massive gold-lip oyster, *P. maxima*, produces correspondingly larger pearls.

Shells

Shells have been employed for practical purposes, for decoration and as a medium of exchange from very early times, and are still largely used in all these ways by savage tribes, and for decorative purposes by civilized peoples. The money cowry (*Cypraea moneta*) has been extensively used in almost every part of the world for purposes of exchange, the shells varying in value according to their size, and other shells have been similarly used in various regions.

R

At the present day certain shells are of considerable commercial value as the source of the mother-of-pearl used for the manufacture of such articles as buttons, studs, knife handles, brooches, fans and all manner of inlaid work. The shells principally employed are those of the gold-lip pearl oyster and the large *Trochus* (a tropical species of the top shells of temperate shores). These are or were important fisheries for the former in the Persian Gulf, the Red Sea, particularly the northern and north-western coasts of Australia with headquarters at Thursday Island in the Torres Strait, and also the Pearl Islands in the Gulf of Panama and other regions in the Indian and Pacific Oceans. The very thick mother-of-pearl which lines the inner surface is cut by machinery and then worked up into the desired shape.

Among the many decorative shells we may again mention that of the abalone (*Haliotis*). When cleaned and polished, this flat and perforated shell is a beautiful object and is often used in shops as a reflector behind electric lights. A New Zealand species, the pawa, blue or green in colour, was much used by the Maoris for decorative purposes. Today all the exotic shells of tropical seas are being collected for sale to amateur conchologists and others with only transitory interest in them, this to an extent that the survival of some is endangered. These shells have become objects of commerce!

Artificial Pearls

Singularly enough, artificial pearls depend for their manufacture upon a marine product. About the middle of the seventeenth century, a French rosary maker called Jaquin discovered that fine flakes of a lustrous, pearl-like substance could be obtained from small freshwater fish. He prepared a thick suspension of this and, by coating alabaster or wax beads with it, succeeded in producing extremely good imitation pearls, and, incidentally, laid the foundations of the modern artificial pearl industry.

The pearly substance, or 'essence d'Orient' as it is called by the French, is really guanin, a waste product like urea, which is found in many fish, though only in a few is it suitable for manufacturing purposes. Usually it is present as a dull powder but for the preparation of pearl essence it needs to be crystalline because only in that form is it lustrous, the minute blade-like crystals reflecting light and breaking it up into the colours of the rainbow. The silvery appearance of the underside of many fish is due to the presence of crystalline guanin in the skin.

In Europe the little freshwater ablette is the main source of pearl essence, in Great Britain the herring, and in America the sardine, herring and other fish. The crystals are extracted by washing the scales and scrubbing them with a mechanical stirrer, the sediment being finely separated in centrifuges which throw the solid matter to the sides and leave the water in the middle. Two types of 'pearls' are made, one from hollow and the other from solid glass beads. The former are coated on the inside with pearl essence and gelatine, the bead being revolved rapidly until a uniform coating is obtained when the cavity is filled in with wax. The more durable solid beads which are made of opaque glass, are given six or more coats of pearl essence usually mixed with celluloid.

For the best imitation pearls the pearl essence has to be chosen very carefully because the effect produced depends on the size of the guanin crystals.

It is comparatively easy to distinguish an imitation from a real pearl. The hollow glass imitation pearls can be detected by the sharp reflection given by the glass surface and by their lightness. The solid variety are not so easy to detect, but the pearly coating can be cut off, is inflammable, and can be dissolved with amyl acetate or acetone, while the coating does not extend evenly up to the edge of the whole through which the string passes. The genuine pearl has a definite weight for its size and is somewhat iridescent; it gives no sharp reflection from its surface and the string hole has clearly been drilled. Acetone and amyl acetate have no effect upon its surface which is, however, unlike those of the imitation varieties, attacked by acid, since the pearl is calcareous.

Precious Coral

The beautiful red coral of commerce (*Corallium rubrum*) (Plate 29) is quite unlike the 'stony' corals which constitute the great mass of coral reefs, being a 'false' coral more nearly allied to the dead-men's-fingers and the sea fan of temperate waters. The red coral substance is not exposed in life but forms the central supporting framework of the colony, on the surface of which is a soft crust through which ramify the canals which connect the flower-like white feeding polyps with which the surface is dotted.

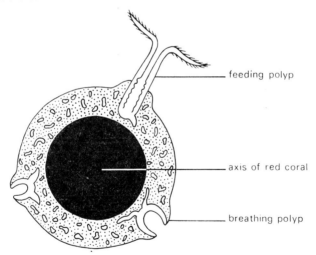

feeding polyp

axis of red coral

breathing polyp

Section showing the structure of a branch of the red coral, *Corallium rubrum*

Red coral is found especially in the Mediterranean, off the south of France and around the coasts of Corsica, Sardinia and Sicily and also along the north coast of Africa from the Straits of Gibraltar to Tunis. It spreads into the Atlantic to some extent and is fished off the Cape Verde Islands. A very similar red coral is found off the coast of Japan.

Coral has been prized by mankind from the very earliest times, not only as an ornament but as a charm against pests, an antidote to poisons and enchantments and as a kind of universal panacea. The story is told in Greek mythology of how after Perseus had slain the Medusa and had thrown her head on the sea shore, the sea-nymphs threw pieces of seaweed at the head and watched them turn into stone. When these stony weeds were washed back into the sea they produced seeds which developed into coral for, as the Greeks had observed, red coral is soft in water but turns hard when exposed to the air.

There appears to have been an extensive trade in coral with the Eastern peoples who valued it as a jewel more highly than the emeralds, rubies and pearls they were willing to give in exchange for it, and this trade penetrated as far as India and China. The early Celtic population of Britain and Ireland also prized coral very highly, probably obtaining it from the Mediterranean by way of Gaul. An interesting evidence of this early trade, which apparently preceded the Roman Conquest, was provided by the discovery, some years back, of a great bronze shield studded with five large pieces of coral, in the bed of the River Witham in Lincolnshire.

Right down to the end of the eighteenth century, red coral held a high place in the esteem of physicians and figured largely in their prescriptions, but the increased knowledge of chemistry and drugs finally showed that red coral was just as valuable a medicine as powdered chalk, of which it principally consists. The only relic of its original virtues which still lingers is the belief that it assists children to cut their teeth, the origin of the still prevalent custom of giving small children necklaces of coral beads.

A black coral is also mentioned in ancient and classical literature. It was also believed to have great medicinal virtues and was found in the Mediterranean and in the Red Sea and Persian Gulf. It seems to have been the black horny skeleton of antipatharians—'false corals' with a bush-like growth. Having no decorative value this coral is no longer used in Europe but bracelets and other articles made of it are still worn in China, Japan, the Malay Archipelago, and throughout the Indian Ocean, among other things as a cure for rheumatism and as a safeguard against drowning.

Turtles

There are only seven species of truly marine turtles, one of them only recently recognized. All are tropical, living purely marine lives with the females alone coming ashore to lay their eggs at the top of sandy beaches. The hatched young emerge to spend years at sea before returning to spawn. Unless carefully protected, which is seldom possible, the eggs are collected in such numbers and the females killed, most species are at risk, the populations of some reduced to dangerous levels. Only two, both present in all tropical seas, are of wide economic importance, the edible green turtle (*Chelonia mydas*) and the hawksbill (*Eretmochelys imbricata*) (Plate 47). The latter is the source of 'tortoise-shell' (never produced by the land living tortoises). The animals are ferocious, armed with powerful jaws—toothless as in all turtles but with horny plates—with which they seize the fish, crustaceans and large molluscs on which they feed. The thick shell plates have to be separated from the underlying bone by heat and are then moulded into the desired shape by immersing in hot water.

The green turtle is highly edible. It was the major source of fresh meat to the explorers of the Gulf of Mexico and the Caribbean and is eaten widely today and the source of turtle soup. It is an inoffensive vegetarian browsing on the roots of eelgrass, the bitten stalks of which rise to the surface and indicate its presence when the turtles can be caught in the bullen, a heavy ring of metal with a deep attached net which is dropped over the feeding turtle and in which it becomes entangled. Both males and females can be caught in this way but only females come ashore during the brief egg-laying period. The eggs, about the size of ping-pong balls and without the breakable shell of a hen's egg, retain condition for a long time and are all too eagerly sought. Recent success by Mariculture Ltd. in breeding turtles on Grand Cayman Island is a major triumph in conservation.

Apart from certain kinds in fresh water, the other turtles most prized as food are the salt-water terrapins which live in salt marshes along the east coast of America from Massachusetts southward as far as South America. These are captured in large numbers for the market, either taken from the marshes during the breeding season or, in the more northern regions, dug from the marshes just as they are beginning their winter sleep or hibernation. At this time they are extremely fat, owing to the reserves of food they have accumulated to carry them over the winter, and so are excellent as food.

Sponges

As mentioned previously sponges are animals. There are many different kinds of sponges but only a few of any commercial importance, and these are to be found in sufficient quantities in certain localities only.

The original sponge fisheries were in the Mediterranean where from remotest antiquity the Greeks of the Ægean Islands have pursued this occupation. Until the middle of the nineteenth century the supply of sponges was derived solely from the Mediterranean, but in 1849 new grounds were discovered off the coasts of Florida and the Bahamas Islands. These two regions are still the sites of the most important sponge fisheries of the world; and of the sponges obtained, those of the Mediterranean are of the best quality. Some of the finest sponge beds in the world lie in Egyptian territorial waters along the North African coast, stretching west from Alexandria for a distance of 483 km to Sollum.

There are three main species of sponges fished in the Mediterranean, the honeycomb, the turkey-cup (Plate 47), and the zimocca. Of these the turkey-cup is the softest in texture. There is also another sponge known as the elephant's-ear sponge, on account of its shape.

The Florida sponges are on the whole of a coarser texture than the Mediterranean sponges. There are several kinds known as sheep's wool, yellow, velvet, grass and glove sponges. The Bahamian and Florida sponges were decimated by a fungal disease in 1938 and 1939 and, nine years later, by some unidentified agency. The industry has since recovered although it is never likely to regain its one-time importance.

Most of the sponges are fished by divers who are most skilled in the Mediterranean where the sponges on the whole grow in deeper water. The sponges were collected either by men in diving dress—machine divers—or by naked divers. The naked divers of the Ægean were unsurpassed for their powers of endurance, and spread from their own islands to fish the north coast of Africa, the central Mediterranean, and some even to the sponge beds in Florida. They fished, in Egyptian waters, chiefly in depths of 22 to 67 m, while they were known to go as deep as 73 and 76 m, generally staying down for about two minutes, though the more expert remained under water for four minutes. The diver plunged into the water grasping in his hand a marble slab, 14 kg in weight, which quickly took him to the bottom. To a ring on his arm was attached his life-line. At a given signal he was hauled up by this as quickly as possible by two leather-gloved men, leaving his marble slab to be pulled up separately by another rope. Modern divers, however, largely use the now ubiquitous aqualung.

The living sponge is black and slimy in appearance owing to the living tissues which cover the skeleton. In order to prepare it for market this living tissue must be removed, for it is only the skeleton which is used. The fresh sponges are laid on deck and stamped upon or beaten; they are then strung together and hung over the side of the ship to macerate for a day. They are once more taken on deck and stamped upon, washed in tubs of sea water, and hung up to dry. At the end of the week they are taken ashore and spread on the sand to bake in the sun, after which they are packed up in sacks ready for sale.

Nowadays artificial plastic sponges have largely replaced the natural sponge for most purposes.

Salt

In many parts of the world salt is obtained from the sea. A thin layer of sea water is allowed to evaporate in the sun and crystals of sea salt are deposited. In order that this may be a commercial proposition it is necessary that the summer should be long and sunny and the climate warm and free from rain. This manufacture of sea salt or 'solar salt', is therefore confined to countries like Italy, Spain, and the coasts of California. The sea water is run into each of a series of shallow ponds in turn. The crystallization of the salts present is not uniform, and as the density of the water changes under evaporation the composition of the salt deposited varies. By running the water through a series of evaporating pools some of the impurities are thus removed, and certain iron salts and calcium carbonate and sulphate have been deposited before the water passes into the main pool in which most of the sodium chloride will crystallize. The finished product, however, still contains impurities and may be refined by recrystallization in fresh water.

In addition to the sodium and chlorine combined to form salt, the sea contains every element, although only in a few cases is extraction economically feasible. The major exceptions are magnesium and bromine where the sea provides a substantial proportion of the world supply.

Products from Seaweeds

There is a great variety of seaweeds growing along rocky shores and many of these are used by man. Around the coasts of the British Isles there are a few species that are eaten. Of these carragheen or Irish moss (*Chondrus crispus*) is eaten in Ireland, dulse (*Rhodymenia palmata*) in Scotland, and laver (*Porphyra laciniata*) in England and Wales. These, all red, weeds are eaten in various ways, laver, for instance, being washed and then boiled for a considerable time with a slight addition of vinegar. After boiling, the laver, which has shrunken to a gelatinous mass, is made up into small cakes, coated in oatmeal, and fried.

The Japanese are by far the greatest consumers of seaweeds which they prepare in a wide diversity of ways. There are restaurants in Tokyo and elsewhere which deal solely with seaweed dishes. Seaweeds are collected from the shore but also off-shore brown laminarians which are torn off the bottom with long hooks. Their most important seaweed, allied to the laver eaten in Wales, the 'red' *Porphyra tenera* or 'nori' is intensely cultivated on bamboo poles arranged in the sea. After harvesting the weed is dried forming dark brown sheets which are then cooked in various ways.

Another food obtained from seaweeds is agar-agar jelly, which is obtained from red varieties. The weeds are boiled and treated until a gelatinous material is produced which can be put to various uses, such as the preparation of soups and gravies, jellies, ice-creams and sweets. It has also a certain medicinal value, and is much used by bacteriologists as it provides a clear, gelatinous medium ideally suited for the growth of cultures of bacteria. Usually imported from Japan, during the second world war it was manufactured in Great Britain from an intertidal weed, *Gigartina stellata*.

The large brown seaweeds, the laminarians or kelp weeds, have a variety of uses. They are a food for sheep in the Orkneys and a valuable manure, rich in potash and free from weeds and pests, especially along the west coasts of Scotland and Ireland. They have a long and interesting history as a source of important chemicals. Beginning in the early years of the eighteenth century, great collections of these weeds were dried in the sun and then burnt in shallow pits, soda and potash being extracted from grey residue or kelp. In Great Britain this industry ranged from Orkney to the Scilly Isles, but with the coming of cheaper sources of these substances the industry declined in the early nineteenth century only to rise again with the discovery of iodine in 1811 and later knowledge of the importance of this in medicine. In France iodine may still be produced from these weeds but not in Great Britain; it is cheaper to import it from Chile where it is produced as a by-product of the nitre industry.

But kelp remains a valuable product because of the contained alginic acid which, with its salts—alginates—provides the strength and elasticity needed for life in the turbulent waters in which these large brown seaweeds live. When extracted these substances have a wide variety of uses in the textile, paper, cosmetic and pharmaceutical industries. The major source comes from the kelp beds off the coast of California. These are composed of *Macrocystis pyrifera*, the largest of seaweeds with stipes, buoyed by floats, extending for lengths of 30 to 300 m. This is mechanically harvested from ocean-going barges. Extraction of alginates in Great Britain is from laminarians with supplies coming largely from the west of Ireland.

Chapter 17

INFLUENCE OF MAN ON MARINE LIFE

From the time that man became a fully conscious social being he began to influence his environment. Over the centuries effects have grown ever greater, in some aspects ever more devastating, as the present condition of many 'developed' countries reveals. These effects were for long almost entirely confined to the land, the construction of even the most primitive collection of dwellings and the simplest beginnings of agriculture immediately influencing the surrounding area. Effects on the sea were much slower to appear. Here man was initially very much of a stranger, concerned only with the sea as a source of food, at first from the shore and very shallow waters, and then more gradually as a means of transport as commerce developed. Possibly his first attempt to modify, at any rate to exploit, the marine environment was the construction of stone fish traps over or around which the tide advanced and behind which fish were retained when the sea retreated. Later development of ports was often accompanied by establishment of semi-stagnant areas of water usually the recipient of much human waste. The problem of pollution had appeared.

For centuries, however, man's effect on the marine environment and on its contained life was of very minor significance. Historically, the problem really becomes noticeable in the nineteenth century and in countries first affected by the Industrial Revolution. The population became concentrated into large towns, sometimes on the coast but always in speedy connection with it by way of the spreading network of railways. To meet the needs of the towns, all manner of fishing activities were intensified, from trawling and drifting for bottom- or surface-dwelling fish to greater search for edible shellfish.

In Chapter 15 something has been said about over-fishing, the effects of which can be countered only by international agreements limiting the intensity of these purely hunting activities. On land such activities passed early into domestication and animal husbandry but this process inevitably takes longer in the sea although such mariculture is already well developed for oysters and mussels (Chapter 16) and there are promising beginnings for both crustaceans and certain fishes. Nevertheless even here irrevocable damage has been done; rich oyster beds, such as those in the Firth of Forth with a former production of many millions annually, have been totally destroyed. Indeed probably no purely natural oyster bed now exists in Europe. It is less easy to destroy complete populations of mobile fishes but the slower breeding marine mammals have suffered terribly at the hands of man. The great sea cow, a sirenian allied to the still precariously surviving manatees and dugong, weighing up to 3 t and living entirely on kelp weed, was exterminated within a generation of its discovery by Steller (after whom it is named) during Bering's second Pacific voyage in 1741.

Whales have suffered almost as badly although possibly saved by greater mobility. The blue whale, the largest animal ever evolved, has been hunted frighteningly near to extinction in the Southern Ocean. In other instances protection has come just in time, notably with the Californian grey whale, particularly vulnerable because breeding is confined to a few lagoons in Baja California. Total numbers fell to a dangerously low figure from which they have now returned and the future of this whale appears secure. Decimation of the originally enormous herds of fur seals which breed on the Pribilov Islands in the Bering Sea was finally stopped by international agreement with future exploitation satisfactorily confined to the taking of excess males not needed for maintenance of the population when active males possess harems of up to 40 females. The delightful sea otter, also of the north Pacific, which was intensely hunted for 200 years as the source of the most valuable of all furs, appeared extinct around the beginning of this century. Here again, following what appeared to be belated measures of protection, the species has re-established itself along the coast of central California as well as farther north off Alaska and the Aleutian Islands. Protection can save threatened populations although in this case action was easier because the animals—inhabitants of the kelp beds—live close inshore and so are largely confined to the territorial waters of the United States.

Man has also affected the sea by his transfer, accidental and deliberate, of marine animals from one ocean or sea to another. The greater number of these introductions, and most of the successful ones, have been accidental. Thus with the introduction into British waters, probably late in the last century, of American 'blue point' oysters for re-laying on local beds, two American pests, the slipper limpet (*Crepidula*) and the oyster drill (*Urosalpinx*) were all too successfully established in this country. With them probably came a boring bivalve (*Petricola pholadiformis*)(the specific name implying its superficial resemblance to the rock-boring piddocks (see p. 102)), which is now common in rocks around the mouth of the Thames and along the coast of Essex. A further competitor of oysters has come from the Antipodes in the form of an acorn barnacle, *Elminius*, which appeared on the south coast of England during the last war, possibly brought by a boat coming at speed direct from Australia or New Zealand, and which has spread widely along southern and western coasts of England. It competes for settling space with oysters because it breeds at the same time as they do, unlike the native barnacles which breed earlier in the year.

A more involved picture of the unpredictable consequences of another such accidental introduction of a foreign species can be seen in the Black Sea. After the war the Russians appear to have introduced Japanese oysters (*Crassostrea gigas*); certainly around 1947 there is the undoubted appearance of the large Japanese oyster drill (*Rapana*) an animal having a shell twice the height and four times the capacity of that of the American *Urosalpinx*. Doubtless this came from Japan in a very early stage or as egg capsules from which the young hatch out.

In any event it soon established itself and proceeded to bore into and consume not only oysters and mussels but also other bivalves that had been the food of bottom-living fish. So much on the debit side; on the other side increasingly large numbers of sand eels (*Ammodytes*) now enter from the Mediterranean to feed on the large stocks of young oyster drills, while the sand eels themselves provide food for another fish, *Sargus*, which had previously fed exclusively on worms and crustaceans. The whole economy of marine life in areas of the Black Sea has thus been altered as a result of the impact of this accidentally introduced species. A further interesting consequence concerns the local hermit crab, a species of *Clibanarius*, previously regarded as a local 'dwarf' race. It had for-

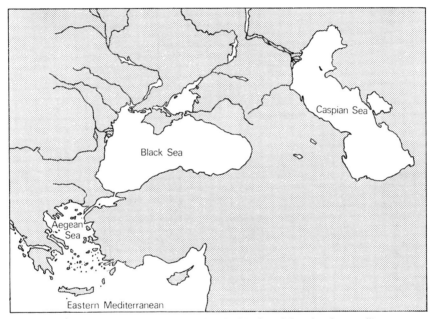

Map of the Eastern Mediterranean with the Black and Caspian Seas to illustrate matters discussed in the text

merly to make do with the small snail shells then alone available, but is now revealed as capable of attaining far larger size within the much more capacious shells of *Rapana*.

Another instance of the devastating effects of accidental introduction concerns the appearance, presumably carried in wooden boats coming from Europe, of the Atlantic shipworm (*Teredo navalis*) in San Francisco Bay where it did catastrophic damage to all wooden erections during 1920. The waters of the Bay have low salinity which is an effective bar to the inward passage of the local shipworm, a species of *Bankia*, but not to this Atlantic species.

Many deliberate attempts have been made to establish foreign, but obviously desirable, species. This has been particularly true of New Zealand which has been all too successful in introducing a host of terrestrial plants and animals. But success has been meagre in the sea. Transplantation of the European lobster and the herring have been attempted, the former on several occasions. But successful accomplishment of the life-history in marine animals depends on conformity with water movements which are most unlikely to be similar in the Antipodes. This is also true of the European salmon although one land-locked race (feeding as well as spawning in fresh water) has been established. A species of Pacific salmon which spends very little time, apart from spawning, in fresh water does appear to have been successfully established. Various species of trout have extensively colonized fresh waters and one has actually left fresh waters to feed in the sea, just as sea trout do in Europe.

However, the most ambitious attempt to transfer a major economic asset from one ocean to another has been made by the Russians. Early attempts to trans-plant eggs of the Pacific chum or dog salmon to the rivers of Murmansk were unsuccessful; later between 1956 and 1959 some 13 million fertilized eggs of the chum salmon and almost four times that number of those of the pink or hump-back salmon were transferred by air to the rivers of the Kola Peninsula in north Russia. Later large numbers of adult pink salmon were caught around these coasts and appreciable numbers farther afield in Norwegian fjords and off Ice-land, with odd captures off the east coast of Scotland and even one in a river in north-west England. This was in 1960; since then few have been caught (outside Russian waters at any rate) except in 1965 when some chum salmon were caught in northern Norway. It does not appear as though the experiment has so far been really successful, although if it were that might be all to the good because these salmon, especially the pink, spend little time in fresh waters. Spawning occurs near the mouths of rivers so that the fish would not compete for food with young Atlantic salmon (parr and smolts) which spend one to three years feeding and growing in fresh waters.

The Russians are also said to be attempting to establish in the north Atlantic the giant north Pacific 'kingcrab', *Paralithodes* (see p. 241). If successful this would introduce an animal of the highest economic importance into Atlantic waters. But, of course, there are the unpredictable consequences that follow introduction of a new species into any balanced ecological system. Striking success was, however, achieved in the Caspian Sea. This is the largest land-locked body of water in the world, five times the area of Lake Superior. It was formerly connected with the sea and so comes within the province of this book. The present salinity is about one-fifth that of ocean water. Its water level, some 30 m below that of the sea, is maintained by a somewhat uncertain balance between the inflowing river water, largely from the Volga, and evaporation. Of recent years, due to withdrawal of river water for irrigation, the level has signi-ficantly dropped. It is inhabited by a somewhat sparse and unbalanced 'relict' fauna, i.e. of animals which have managed to survive the gradual change in conditions since it lost connection with the sea. Unlike the fully balanced condi-tions in the sea, certain ecological 'niches' are unfilled and this became apparent to the Russians who were concerned about the fall of productivity in the 1930's. They decided to remedy the major gap in the ecological balance by introducing a species of *Nereis*, a polychaete worm which could live in the wide areas of largely unpopulated mud and there feed on the contained organic matter. This introduction took place between 1939 and 1941 and only a few years later these worms began to be found in the stomachs of sturgeon, the most important local fish, its eggs eaten as caviare. By 1948 the worms were present in an area of 30,000 km^2 and were the third most important bottom-living invertebrates. They burrow in the mud, emerging at the end of a one-year life to spawn in the water above when they are consumed in vast numbers not only by fish but also by birds who congregate round the shores where the worms, drained of sexual products and dead, are cast up in great numbers. In brief, a route has been found and exploited for the passage of the rich stores of organic matter in the mud into the bodies of fish and of birds.

Pollution is another man-made problem although it should initially be realized that, unaided, the sea itself may 'go bad'. Over-production of phytoplankton (due to excess of nutrients) causes devastating 'red tides', especially frequent off the west of Florida, while occasional deflection of warm currents south along the coast of Peru leads to mass mortality in the teeming life of the Humboldt Current, even the birds starving to death.

Broadly speaking, pollution comes from domestic sewage, from industrial waste, from pesticides or sometimes from heated water effluents from power stations; it also comes from discharge or escape of crude oil. All but the last primarily affect rivers and estuaries although there is continual discharge and slow build-up in the sea. The persistent DDT has even been found in the bodies of Antarctic birds, passing through so many other organisms to get there. It is now quite possible to prevent such pollution by passing sewage and industrial effluents through suitable cleansing processes while all pesticides used should be capable of quick breakdown, be, as we now say, 'bio-degradable'.

A classic instance of pollution from largely domestic sources is displayed in the harbour of the town of Tunis where sewage from a quarter of a million people is discharged. This abnormal addition of nutrient salts leads to an equally excessive production of plant life, largely green seaweeds, which then die and rot with the usual effects of mass organic decay. This and similar problems can be solved by sewage treatment or, by suitable engineering, production of a free flow of water through the harbour. But both require heavy expenditure and inevitably must wait.

Discharge of crude oil into the sea, of which the *Torrey Canyon* episode and the underwater 'leak' off Santa Barbara will long be the classic events, has the effect of liberating vast quantities of a viscous fluid which will inevitably strike an adjacent coast, precisely where and when depending on the prevailing winds. The first detergents used, while certainly dispersing the oil, had devastating effects on shore animals and plants. Such detergents, and certainly none in the quantities initially used, are unlikely to be employed again. It is probably simpler to sink the oil by adding slaked lime or sawdust and allowing it to settle on the sea bed where it seems to do little damage and gradually to be broken down by bacteria—because oil is 'bio-degradable'. It is just the sheer amounts needing attention that may be daunting while sea birds receive devastating damage from this source. Fortunately a great deal has been learnt since the early major oil spills.

When discussing the problems of Tunis harbour, the need for engineering was stressed. And it is in his capacity as an engineer, with the ever greater powers available in impressive earth-moving equipment, that man is now making his greatest impact on the marine environment. This, of course, began as we have indicated with construction of the first simple harbours and with that of any obstacle to the free movement of local seas.

Modern works of civil engineering are large enough to affect whole countries. One in Northern Europe is the enclosure of the Zuider Zee, completed with the closure of the dyke in June, 1932, when a wide area of saline water was cut off from the sea gradually to be converted into a freshwater lake. The Zuider Zee had been formed by a great incursion of the sea in the thirteenth century following which a marine population had gradually developed in the somewhat brackish waters. This included food fishes such as herring, anchovy and plaice which were the mainstay of fishing communities which became established on the banks. A colony of seals settled itself on the island of Urk. Prior to the building of the dyke and the formation of the 'polders' which have added materially to the area of the Netherlands, the Dutch made a thorough biological survey so that the effect of the gradual change into the finally purely freshwater Ijsselmeer could be accurately assessed. As was to be expected, the strictly marine species soon disappeared, but some estuarine species—found normally only in partially saline water, never in entirely fresh or fully saline waters—survived. These included an interesting endemic crab, *Heteropanope tridentata*, and a species of acorn barnacle together forming another 'relict' fauna reminiscent of the days when there was open communication with the sea. For the rest, as salinity declined the new lake was gradually dominated by freshwater animals, largely insects, molluscs and fish such as carp, which migrated into it from the rivers which flow into the Ijesselmeer or Yssel Lake.

This was a gradual change, and the only ensuing problem was a temporary plague of harlequin flies (*Chironomus*), the larvæ of which live in fresh water, which emerged in such vast numbers as even to block the radiators of cars on the road along the dyke. They established themselves before the freshwater fish which, coming later, proceeded to feed on the insect larvæ and so controlled their population.

A sudden change of conditions, from fully saline to completely fresh water, occurred during the recent formation of the Plover Cove reservoir in the New Territory of Hong Kong. This was initially an arm of the sea some 5 km long and half as wide which was then cut off by the erection of dams. The sea water was then pumped out while fishermen were allowed to catch everything they could. Finally everything possible in the way of living or recently living matter was removed from the small remaining quantity of sea water and from the largely mud bottom. The basin was then allowed to refill with fresh water, partly from the surrounding catchment area, partly by diverting water through the mountainside from other areas. Finally what appeared superficially to be the former conditions returned. Plover Cove was again full of water but actually to a higher

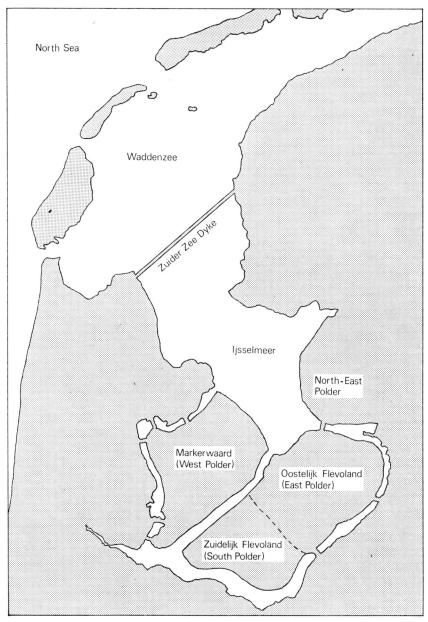

North Sea

Waddenzee

Zuider Zee Dyke

Ijsselmeer

North-East
Polder

Markerwaard
(West Polder)

Oostelijk Flevoland
(East Polder)

Zuidelijk Flevoland
(South Polder)

Map of the former Zuider Zee showing the dyke which converted its saline
waters into the freshwater Ijsselmeer with much of its former extent drained
and converted into polders

level than before owing to the height of the dams. Anticipated danger from accumulation of decaying organic matter with consequent deoxygenation of the water (as occurred in the great African man-made lakes at Kariba and Volta) did not materialize, partly because so much animal life had been removed before fresh water entered, and partly because the long axis of the new reservoir was in the direction of the prevailing winds so that there was a continual surface flow with consequent upwelling of water from below; the water was thus kept well mixed with adequate oxygen throughout. The marine population entirely disappeared and has been replaced by one coming from the small streams that originally flowed into the sea.

The greatest of such engineering feats, one indeed which united the two great ocean basins, the Indo-Pacific and the Atlantic, with largely distinct populations, was completed over a century ago. The Suez Canal was cut through the mostly sandy area between the northern end of the Gulf of Suez and the Mediterranean, opening only a few kilometres east of the Nile delta. There were no freshwater locks such as those at Panama which interpose an almost completely effective barrier between the Atlantic and Pacific ends of the canal. At first sight therefore it might appear that the very different marine populations at the two ends of the canal would soon become thoroughly mixed. It is interesting to relate how this did not immediately happen and why this process has taken so long to occur and has been largely a one-way movement. There are also the complicating effects of another major engineering project affecting the same area of the Mediterranean, namely the Aswan High Dam.

The Suez Canal is 162 km long from Port Said in the north to Suez in the south. Owing to forces generated by the rotation of the earth, sea level is some 5 to 8 cm higher at Suez so that a current normally flows from south to north through the canal. However, during the period July to September the current is reversed, owing to the piling of water in that corner of the Mediterranean due to the then prevailing north-west winds, reinforced until recently by the outflow of water from the Nile. Until the completion of the first stage of the Aswan Dam in 1966 this was the period of the Nile floods when enormous volumes of water poured out, both lowering the salinity and raising the level of the sea. On balance one would therefore expect a considerable passage of animals from the Red Sea into the Mediterranean with less in the opposite direction. However, other factors come into the picture. A major initial barrier to passage in either direction was the presence of the Bitter Lakes which, as shown in the map, represent about a fifth of the length of the canal. A layer of salt formed by the drying up of a former extension of the Gulf of Suez and of a calculated weight of around 97 million tonnes covered the bottom of the Great Bitter Lake when the canal was opened in 1869. Dissolution of this raised the salinity to about double that of the open sea, conditions lethal to almost every marine animal. Further obstacles to passage are the soft mud of the canal bottom and also the turbid water produced by passing ships.

When the fauna of the canal area was surveyed by a British expedition in 1924, 55 years after the two oceans were united, only fifteen Red Sea species were found in the Mediterranean while nine had made passage southward. Today, over a century after the canal was opened, conditions have changed and animals are passing freely through the canal. This is primarily due to dissolution and removal by the slowly moving canal waters of the bulk of the salt deposits. Salinity has gradually been reduced and no longer forms a barrier. Moreover, the

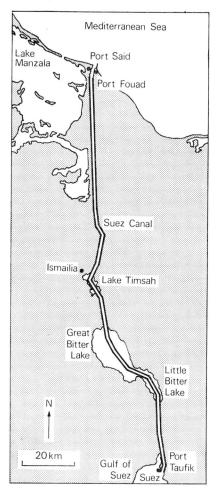

Map of the Suez Canal showing how it connects the Atlantic, by way of the Mediterranean, with the Indo-Pacific at Suez, also the position of the Bitter Lakes, the high salinity of which, until recently, prevented the passage of animals through the canal and consequent mixtures of the populations of these two great oceanic regions

canal was dredged deeper in 1961 to 1963 so that the flow of water has increased by 50%. At the same time stoppage of the Nile floods means that low salinity no longer impedes southward migration of animals.

The most recent surveys reveal the presence of at least 114 Indo-Pacific species in the south-east Mediterranean, with 16 Mediterranean species in the Gulf of Suez. The former consist of 2 sponges, 2 hydroids, 10 polychaete and other worms, over 40 crustaceans, four of which are planktonic and the bulk of the remainder shrimps, prawns and crabs, 20 molluscs, 4 echinoderms (starfish, brittlestars and sea urchins), 6 sea squirts and 26 fishes. Three sponges, 2 sipunculid worms, 1 crustacean and 8 fishes have passed through in the opposite direction. Two Indo-Pacific seaweeds have made the passage north and the same number of Atlantic species has moved into the Red Sea.

The greater northward movement is certainly due to the prevailing flow in that direction. Certain animals, notably fish and many of the larger crustaceans will have migrated as adults, but the majority of the invertebrates must have done so during the limited period of up to perhaps two weeks when, as larvæ, they are members of the plankton and can be carried passively in the prevailing water currents. This is particularly probable for the echinoderms, which could not proceed as adults through the soft mud, and also for most of the worms and molluscs. But, of course, such migration is only possible if spawning occurs during the period of flow, north or south as the case may be, through the canal. Even when safely through, larvæ may be unable to find conditions suitable for settlement and metamorphosis. This is certainly true for animals living attached to rocks in the Gulf of Suez because similar conditions do not exist within hundreds of kilometres of Port Said. Hence passage of the larvæ of such animals can have no permanent effect on the Mediterranean fauna.

These mixings of the populations of the two great ocean systems are fascinating biologically; new ecological systems are inevitably established as the more fitted species oust the less fitted and new feeding possibilities are exposed and exploited. But there is also the economic side of things. Here one must distinguish between the advent of new species in the south-east Mediterranean and the effect of the stoppage of the Nile floods. The former has certainly been beneficial, some useful food fishes have come into the Mediterranean, also the valuable penaeid shrimp (*Penaeus japonicus*) the one cultivated in Japan. Pearl oysters have also entered. Both fish and prawns, the latter forming a new fishery, have certainly increased the variety of food from this essentially impoverished area. Or rather they may have helped to offset the major, almost catastrophic, reduction in productivity due to cessation of the annual Nile floods. These formerly not only enriched the agricultural areas on either bank of the river, they finished by adding enormously to the nutrient salts in the Mediterranean. The result of stoppage has been seen most dramatically in the yield of the sardine fisheries, particularly along the coast of Israel, which have sunk to a mere fraction of their former size. So the final result of these two major engineering projects would appear as a great enrichment of the Mediterranean fauna by animals coming from the Indo-Pacific but with a much reduced overall productivity. The new species may in some cases be more valuable economically than indigenous ones, but total productivity is not for that reason increased; this depends on the local supply of nutrient salts which here, so remote from the major ocean systems and at the extremity of what is at best the poorly productive Mediterranean, came from the annual bounty from the Nile, from the mountains, plains and swamps of Abyssinia and of central Africa.

What has happened at the two ends of the Suez Canal has naturally been the concern of those, especially in the United States, who are interested in the possibilities, now being actively canvassed, of a sea-level canal across central

America. This would not involve passage through what have been described by Dr. I. Rubinoff as the 'highly restrictive filter' of the freshwater locks in the Panama Canal. Even with the most careful precautions, passage of animals through a purely marine canal, like Suez without any Bitter Lake, would be inevitable. The problem presented is again that of a mixing of the Atlantic and Indo-Pacific fauna but this time at the other end of their range. Here the Atlantic fauna is far more vigorous than in the Mediterranean and might have as profound an effect on the fauna of the eastern Pacific as animals from there might have in the Caribbean. Since time is available, it is possible to test out possibilities by studying the interactions of animals of similar habits from the two oceans and, most immediately important, the vulnerability of animals from the one side to the parasites or predators from the other. An animal such as the crown-of-thorns starfish (*Acanthaster*) introduced into the Caribbean could inflict devastating damage on Atlantic reef-building corals which at present have no significant enemies.

Other grim possibilities loom ahead. Apart from the increasing effects of world-wide pollution, some of them at least partially due to engineering projects, there are the possible effects of further reduction or diversion of water for hydro-electric purposes or for irrigation. Possibly the greatest danger is in the Black Sea. There, as described on p. 159, only the surface waters are adequately oxygenated, in deep water oxygen is increasingly displaced by sulphuretted hydrogen and life, apart from the sulphur bacteria which produce this, is impossible. This condition is maintained because the surface waters are less saline than those deeper down and so lighter and there is no mixing of deep and surface waters. But this depends on the great inflow of water from the rivers of southern Russia. This is now being increasingly diverted and if salinity in the surface waters of the Black Sea rises above a critical figure the whole water mass may 'turn over' bringing the foul-smelling, deoxygenated waters to the surface with devastating effects on all life in shallow waters. In dealing with the sea and its life, there is a limit beyond which man cannot pass with impunity.

BIBLIOGRAPHY

Although by no means exhaustive, this list of literature, completely revised for the present edition, will be found useful in relation to the scope of this volume.

General works

ALCOCK, A., *A Naturalist in Indian Seas*, John Murray, 1902.
BARNES, H. (Ed.), *Oceanography and Marine Biology*, George Allen & Unwin Ltd, 1959.
BEEBE, William, *The Arcturus Adventure*, Putnam & Co. Ltd, 1926;
  Half-Mile Down, Harcourt Brace & Co., N.Y., 1934.
BRUUN, A. F. *et al.*, *The Galathea Deep-Sea Expedition 1950–52*, George Allen & Unwin Ltd, 1956.
CARRINGTON, Richard, *A Biography of the Sea*, Chatto & Windus, 1960.
CARSON, Rachel, *The Sea Around Us*, Panther, 1969.
COLMAN, John S., *The Sea and its Mysteries*, G. Bell & Sons Ltd, 1950.
COUSTEAU, J. Y. with DUGAN, J., *The Living Sea*, Hamish Hamilton, 1963;
  and DUMAS, Frédéric, *The Silent World*, Hamish Hamilton, 1953.
DEACON, G. E. R. (Ed.), *Oceans*, Paul Hamlyn, 1968.
DIETRICH, Gunter, *General Oceanography*, Interscience, 1963.
EKMAN, Sven, *Zoogeography of the Sea*, Sidgwick & Jackson Ltd, 1967.
HASS, Hans, *We Come from the Sea*, Jarrolds, 1958.
HERRING, P. J. and CLARKE, M. R. (Ed.), *Deep Oceans*, Arthur Barker Ltd, 1971.
IDYLL, C. P., *Abyss: The Deep Sea and Creatures that Live in it*, Constable, 1965.
MARSHALL, N. B., *Aspects of Deep Sea Biology*, Hutchinson, 2nd impression, 1958.
MILLER, R. C., *The Sea*, Random House, New York, 1966.
MOORE, H. B., *Marine Ecology*, John Wiley & Sons, 1950.
MURRAY, Sir J. and HJORT, Prof. J., *The Depths of the Ocean*, Macmillan, 1912. (Wheldon, 1964.)
NICOL, J. A. Colin, *Biology of Marine Animals*, Sir Isaac Pitman & Sons Ltd, 1968.
OMMANNEY, F. D., *The Shoals of Capricorn*, Longman, 1952.
PICCARD, Jacques and DIETZ, Robert S., *Seven Miles Down. The Story of the Bathyscaphe Trieste*, Longman, 1962.
SHEPARD, Francis P., *The Earth Beneath the Sea*, Johns Hopkins Press, 1968.
STEERS, J. A., *The Sea Coast* (New Naturalist), Collins, 1953.
STEINBECK, John, *The Log from the Sea of Cortez*, Heinemann, 1958.
SVERDRUP, H. U., JOHNSON, Martin W. and FLEMING, R. H., *The Oceans*, Prentice-Hall, 1942.
THORSON, Gunnar, *Life in the Sea*, World University Library, Weidenfeld & Nicolson, 1972.
ZENKEVITCH, L., *Biology of the Seas of the U.S.S.R.*, George Allen & Unwin Ltd, 1963.

The seashore and shallow waters

BARRETT, John H. and YONGE, C. M., *Pocket Guide to the Sea Shore*, Collins, 1958.
BENNETT, I., *The Fringe of the Sea*, Angus and Robertson, 1967.
BERRILL, N. J., *The Living Tide*, Gollancz, 1951;
  The Tunicata, Ray Society, 1950.
BINYON, John, *Physiology of Echinoderms*, Pergamon Press, 1973.
CALMAN, W. T., *The Life of Crustacea*, Methuen & Co., Ltd, 1911.
CARSON, Rachel, *The Edge of the Sea*, Staples, 1955.

CLAYTON, Joan M., *The Living Seashore*, Frederick Warne & Co. Ltd., 1974.
DAKIN, W. J., *Australian Seashores*, Angus and Robertson, 1960.
DALES, R. Phillips, *Annelids*, Hutchinson, 1967.
DICKINSON, Carola I., *British Seaweeds*, Eyre & Spottiswoode, 1963.
EALES, N. B., *The Littoral Fauna of Great Britain*, Cambridge University Press, 3rd edition, 1961.
FRETTER, Vera and GRAHAM, Alistair, *British Prosobranch Molluscs*, Ray Society, 1962.
GRAHAM, Alistair, *British Prosobranchs*, Academic Press, 1971.
MacGINITIE, G. E. and N., *Natural History of Marine Animals*, McGraw-Hill, 1968.
McMILLAN, N. F., *British Shells*, Warne, 1968.
MORTON, J. and MILLER, M. (Ed.), *The New Zealand Sea Shore*, Collins, 1968.
MORTON, J. E., *Molluscs*, Hutchinson, 1967.
NICHOLS, D., *Echinoderms*, Hutchinson, 1969.
NILSSON, L. and JAGERSTEN, G., *Life in the Sea*, Foulis, 1961.
RICKETTS, E. F. and CALVIN, J., *Between Pacific Tides* (4th edition revised by HEDGPETH, J. W.), Stanford University Press, 1961.
STEERS, J. A., *The Sea Coast* (New Naturalist), Collins, 1953.
STEPHENSON, T. A., *The British Sea Anemones* (Vols. I and II), Ray Society, 1935.
TEBBLE, N., *British Bivalve Seashells*, British Museum (Natural History), 1966.
WILSON, Douglas P., *Life of the Shore and Shallow Sea*, Ivor Nicholson & Watson, 1935.
YONGE, C. M., *The Sea Shore* (New Naturalist), Collins, 1949.
YONGE, C. M. and THOMPSON, T. E., *Living Marine Molluscs*, Collins, 1975.

Fish and fisheries

BALLS, R., *Fish Capture*, Edward Arnold Ltd, 1961.
BERTIN, L., *Eels, a Biological Study*, Cleaver-Hume Press Ltd, 1956.
BORGSTROM, Georg and HEIGHWAY, Arthur J. (Ed.), *Atlantic Ocean Fisheries*, Fishing News (Books) Ltd, 1961.
CULLEY, Michael, *The Pilchard. Biology and Exploitation*, Pergamon Press, 1973.
CUSHING, David, *The Detection of Fish*, Pergamon Press, 1973.
FORD, E., *The Nation's Sea Fish Supply*, Edward Arnold Ltd, 1937.
GRAHAM, Michael, *The Fish Gate*, Faber & Faber, 1943.
GRAHAM, M. (Ed.), *Sea Fisheries: Their Investigation in the United Kingdom*, Edward Arnold Ltd, 1956.
HARDY, Sir Alister, *The Open Sea: Its Natural History* (Part II), *Fish and Fisheries* (New Naturalist), Collins, 1956.
HICKLING, C. F., *The Hake and the Hake Fishery*, Edward Arnold Ltd, 1935.
HODGSON, W. C., *The Herring and Its Fishery*, Routledge & Kegan Paul, 1957.
HJUL, Peter (Ed.), *The Stern Trawler*, Fishing News (Books) Ltd, 1972.
JONES, F. R. Harden, *Fish Migration*, Edward Arnold, 1968.
KRISTJONSSON, Hilmar (Ed.), *Modern Fishing Gear of the World*, Vols. I & II, Fishing News (Books) Ltd, 1959 and 1964.
JENKINS, J. Travis, *The Fishes of the British Isles*, Frederick Warne & Co. Ltd (revised edition by WHEELER, A., in preparation);
The Sea Fisheries, Constable & Co. Ltd, 1920.
MARSHALL, N. B., *The Life of Fishes*, Weidenfeld & Nicolson, 1965.
MILLS, D., *Salmon and Trout*, Oliver & Boyd, 1971.
MORGAN, R., *World Sea Fisheries*, Methuen & Co. Ltd, 1956.
NIKOLSKY, G. V., *The Ecology of Fishes*, Academic Press, 1962.
NETBOY, A., *The Atlantic Salmon: Vanishing Species*, Faber & Faber, 1968.
NORMAN, J. R. and FRASER, F. C., *Giant Fishes, Whales and Dolphins*, Putnam & Co. Ltd, 1948;
(2nd Edition by GREENWOOD, P. H.), *A History of Fishes*, Ernest Benn Ltd, 1963.
RUSSELL, E. S., *The Overfishing Problem*, Cambridge University Press, 1942.

RUSSELL, Sir F. S., *The Eggs and Planktonic Stages of British Marine Fishes*, Academic Press, 1975.
SINGER, Burns, *Living Silver*, Secker & Warburg, 1957.
VILLIERS, A., *The Quest of the Schooner 'Argus'*, Hodder & Stoughton, 1951.
WHEELER, Alwyne, *The Fishes of the British Isles and North-west Europe*, Macmillan & Co. Ltd, 1969.

Whales, whaling, seals and turtles

BUDKER, P., *Whales and Whaling*, Harrap, 1958.
BUSTARD, Robert, *Sea Turtles, Their Natural History and Conservation*, Collins, 1972.
CARR, Archie, *The Turtle*, Cassell, 1968.
HARDY, Sir Alister, *Great Waters*, Collins, 1967.
HARRISON, Richard J. and KING, Judith, *Marine Mammals*, Hutchinson, 1968.
HEWER, H. R., *British Seals*, Collins, 1974.
KELLOG, W. N., *Porpoises and Sonar*, University of Chicago Press, 1961.
MATTHEWS, L. H. (Ed.), *The Whale*, George Allen & Unwin Ltd, 1969.
OMMANNEY, J. D., *Lost Leviathan. Whales and Whaling*, Hutchinson, 1971;
    South Latitude, Longman, 1938.
SLIJPER, E. J., *Whales*, Hutchinson, 1962.

Plankton

BEEBE, William, *The Arcturus Adventure*, Putnam & Co. Ltd, 1926.
FRASER, J. H., *Nature Adrift*, Foulis, 1962.
HARDY, Sir Alister, *The Open Sea: Its Natural History. The World of Plankton* (New Naturalist), Collins, 1971.
LEBOUR, M. V., *The Dinoflagellates of Northern Seas*, Marine Biological Association, Plymouth, 1925;
    The Planktonic Diatoms of Northern Seas, Ray Society, 1930.
MARSHALL, S. M., and ORR, A. P., *The Biology of a Marine Copepod Calanus finmarchicus* (Gunnerus), Oliver & Boyd, 1955.
NEWELL, G. E. and R. C., *Marine Plankton. A Practical Guide*, Hutchinson Educational, 1963.
ORR, A. P. and MARSHALL, S. M., *The Fertile Sea*, Fishing News (Books) Ltd, London, 1969.
RAYMONT, J. E. G., *Plankton and Productivity in the Oceans*, Pergamon Press, 1963.
RUSSELL, Sir F. S., *The Medusae of the British Isles* (Vols. I and II), Cambridge University Press, 1953 and 1970.
WICKSTEAD, John H., *An Introduction to the Study of Tropical Plankton*, Hutchinson (Tropical Monographs), 1965.
WIMPENNY, R. S., *The Plankton of the Sea*, Faber & Faber, 1966.

Sea water and hydrography

BARNES, H. (Ed.), *Oceanography and Marine Biology*, George Allen & Unwin, 1959.
BORGSTROM, Georg and HEIGHWAY, Arthur J. (Ed.), *Atlantic Ocean Fisheries*, Fishing News (Books) Ltd, 1961.
CORNISH, Vaughan, *Ocean Waves and Kindred Geophysical Phenomena*, Cambridge University Press, 1934.
HARVEY, H. W., *Biological Chemistry and Physics of Sea Water*, Cambridge University Press, 1957;
    The Chemistry and Fertility of Sea Waters, Cambridge University Press, 1945.
RITCHIE, G. S., *The Admiralty Chart*, Hollis & Carter, 1967.
RUSSELL, R. C. H. and MACMILLAN, D. H., *Waves and Tides*, Hutchinson, 1952.

Corals

BENNETT, I., *The Great Barrier Reef*, Frederick Warne & Co. Ltd., 1973.
CLARE, P., *The Struggle for the Great Barrier Reef*, Collins, 1971.

DARWIN, Charles, *Coral Reefs*, John Murray, 3rd edition 1889.
GARDINER, J. Stanley, *Coral Reefs and Atolls*, Macmillan & Co. Ltd, 1931.
GILLETT, Keith and MCNEILL, Frank, *The Great Barrier Reef and Adjacent Islands*, Coral Press Pty. Ltd, Sydney, 1959, reprinted with revisions, 1967.
HICKSON, S. J., *An Introduction to the Study of Recent Corals*, Manchester University Press, 1924.
SMITH, F. G. Walton, *Atlantic Reef Corals*, University of Miami Press, revised edition, 1971.
WIENS, H. J., *Atoll and Environment*, Yale University Press, 1962.
WOOD-JONES, F., *Coral and Atolls*, Lovell Reeve & Co. Ltd, 1912.
YONGE, C. M., *A Year on the Great Barrier Reef*, Putnam & Co, Ltd, 1930.

Products of the sea

DUDDINGTON, C. L., *Seaweeds and Other Algae*, Faber & Faber, 1966.
FIRTH, F. E. (Ed.), *The Encyclopedia of Marine Resources*, Van Nostrand Reinhold Co, New York, 1969.
HICKLING, C. F., *Fish Culture*, Faber & Faber, revised edition, 1971.
HORSFIELD, B. and STONE, P. Bennet, *The Great Ocean Business*, Hodder & Stoughton, 1972.
IVERSON, E. S., *Farming the Edge of the Sea*, Fishing News (Books) Ltd, 1968.
LOFTAS, T., *The Last Resource: Man's Exploitation of the Sea*, Hamish Hamilton, 1969.
NEWTON, Lily, *et al.*, *Sea-weed Utilization*, Low, 1951.
WALFORD, Lionel A., *Living Resources of the Sea*, Ronald Press Co., 1958.
YONGE, C. M., *Oysters* (New Naturalist), Collins, 2nd edition, 1966.

Pollution

MARX, Wesley, *The Frail Ocean*, Ballantine Books, New York, 1971.
SMITH, J. E. (Ed.), '*Torrey Canyon*' *Pollution and Marine Life*, Cambridge University Press, 1968.

Historical

DEACON, Margaret, *Scientists and the Sea, 1650–1900*, Academic Press, 1971.
HERDMAN, Sir William, *Founders of Oceanography*, Edward Arnold Ltd, 1923.
HUXLEY, Sir Julian S. (Ed.), *T. H. Huxley's Diary of the Voyage of H.M.S. 'Rattlesnake'*, Chatto & Windus, 1935.
MOSELEY, H. N., *Notes by a Naturalist on the 'Challenger'*, Macmillan & Co. Ltd, 1892.
MURRAY, Sir J. and HJORT, Prof. J., *The Depths of the Ocean*, Macmillan & Co. Ltd, 1912.
THOMSON, Sir Wyville, *The Depths of the Sea*, Macmillan & Co. Ltd, 1873.

INDEX

Numbers in *italics* indicate diagrams in the text

T